The Human Howl: Why We Sing and Why We Can't Stop

D. M. Gross

Published by Dustin Gross, 2025.

THE HUMAN HOWL: WHY WE SING AND WHY WE CAN'T STOP

First edition. November 26, 2025.

ISBN: 979-8232625726

Written by D. M. Gross.

Table of Contents

Introduction

The Universal Glitch

Stand in the center of a football stadium during a playoff game. Close your eyes. Ignore the scoreboard, the overpriced beer, and the commercial breaks. Listen to the sound. Fifty thousand people are screaming in unison. They are hitting a specific pitch, holding a specific rhythm, and modulating their volume based on a shared, invisible cue. The sound is physical. It vibrates in your chest cavity. It registers on seismographs. If you strip away the jerseys and the concrete, you are left with a massive, coordinated organism howling at the sky.

This behavior is a biological glitch.

By all standard metrics of evolutionary efficiency, singing is a waste of time. It burns calories that could be used for hunting or foraging. It requires a respiratory system that is dangerously modified, leaving us vulnerable to choking on our own food. It is loud, which in the wild is synonymous with "here I am, come eat me." A silent primate is a safe primate. A loud primate is a target. Yet, humans are the only species on the planet that organizes its entire social structure around the production of complex, rhythmic, melodic noise. We do it at birthdays. We do it at funerals. We do it in cars, in showers, and in places of worship. We cannot seem to stop.

The question is not how we sing. We know how. We push air from the lungs through the trachea, vibrating two small flaps of mucous membrane called the vocal folds, and shape the sound with our tongue and lips. The mechanics are simple. The question is why. Why did natural selection, a ruthless editor that trims away any trait not essential for survival, keep this one? Why did it prioritize the ability to hold a tune over the ability to swallow without risk of death?

The answer is that singing is not a hobby. It is not a decorative art form reserved for the talented few. It is a survival strategy. It is the primary technology our species developed to bond groups, regulate nervous systems, and transmit information before the invention of the written word. We are biologically engineered amplifiers. We were built to howl.

Somewhere along the line, we forgot this. We decided that singing was a performance, a commodity to be bought and sold. We created a binary world of "singers" and "listeners." If you cannot hit a high C with perfect vibrato, you are told to be quiet. You are told you are "tone deaf." You are told to mouth the words. This cultural shift has created a silence epidemic. We consume more music than any generation in history, streaming it into our ears for hours a day, but we make less of it than ever before. We have outsourced our own biological regulation to professionals.

This silence is making us sick. It is isolating us. It is denying us the chemical wash of endorphins and oxytocin that our ancestors relied on to tolerate the misery of existence. This book is the argument for opening your mouth and making noise, regardless of what it sounds like.

Consider the sheer physical improbability of the human voice. To produce a sung note, the brain must coordinate over one hundred different muscles. The diaphragm descends, creating a vacuum that pulls air into the lungs. The abdominal muscles engage to pressurize that air. The vocal folds, which are smaller than a postage stamp, must come together with microscopic precision. If they are too loose, the sound is breathy. If they are too tight, the sound is strangled. They vibrate hundreds of times per second. For a high soprano note, those folds are colliding 1,000 times every second.

Then the sound hits the resonator. The throat, the mouth, the sinus cavities. The tongue, a massive muscle rooted in the throat, must dance out of the way to let the wave pass while simultaneously shaping vowels. The soft palate raises to close off the nose, or lowers to let the sound hum. All of this happens in milliseconds. It happens without conscious thought. You decide to sing a note, and the body executes a complex ballistic maneuver that rivals the coordination required to throw a fastball.

No other animal does this. Birds sing, yes. Whales sing. But their songs are largely hard-wired or strictly functional for territory and mating. They do not gather in groups of thousands to synchronize their breathing and heart rates through a shared melody. Chimpanzees, our closest genetic relatives, cannot do it. They can hoot and scream, but they cannot keep a beat. If you tap a rhythm for a chimp, it cannot tap it back. It lacks the neural connection between the auditory cortex and the motor cortex. Humans are the only ones who can hear a rhythm and physically synchronize their body to it. This ability, called entrainment, is the foundation of civilization.

If you cannot synchronize with the group, you cannot hunt together. You cannot build together. You cannot fight together. The tribe that sings together moves together. The tribe that falls silent falls apart.

This utility explains the risk. The human larynx is positioned lower in the throat than in any other primate. In a chimp, the larynx sits high up, locking into the nasal cavity. This creates a sealed tube. A chimp can drink water and breathe at the same time. It is physically impossible for a chimp to choke to death on a piece of fruit. The food goes down the esophagus, the air goes down the trachea, and the two paths never cross.

In humans, the larynx dropped. Over hundreds of thousands of years, it slid down the throat. This created a large resonant chamber—the pharynx—above the vocal cords. This chamber allows us to produce the vowels a, i, and u. It allows us to speak and sing with nuance. But it also means that the opening to our lungs sits right next to the opening to our stomach. Every time we swallow, we gamble. We rely on a small flap of cartilage, the epiglottis, to snap shut and cover the airway. If it misses, we die.

Nature does not make mistakes like this. Nature does not introduce a fatal design flaw unless the payoff is massive. The ability to produce complex sound was so valuable to the survival of the human species that it was worth the risk of choking. We traded safety for a song.

That trade-off implies that the voice is not just for communication. It is for regulation. When we make a sustained vocal sound, we stimulate the vagus nerve. This nerve, the longest of the cranial nerves, wanders from the brainstem down through the neck, passing the vocal cords, and into the chest and abdomen. It innervates the heart, the lungs, and the gut. It is the command center for the parasympathetic nervous system—the "rest and digest" mode.

When you are stressed, your body is in sympathetic mode. Fight or flight. Heart rate up. Breath shallow. Cortisol flooding the blood. This was useful when we were running from lions. It is less useful when we are sitting in traffic or worrying about a mortgage. Chronic stress destroys the body. It causes inflammation, heart disease, and anxiety.

Singing is the manual override. When you sing, especially long, sustained notes, you are physically massaging the vagus nerve. You are forcing your exhalation to be longer than your inhalation. This breathing pattern signals the brain that the threat is over. You cannot sing a ballad while running for your life. Therefore, if you are singing, you must be safe. The brain gets the message and lowers the heart rate. It dumps oxytocin and endorphins into the system. It metabolizes the cortisol.

This is not a placebo effect. It is mechanics. A mother humming to a crying infant is not just offering comfort; she is using acoustic vibration to externally regulate the child's nervous system. The baby's heart rate synchronizes with the mother's voice. The vibration travels through the chest wall and settles the infant's breathing. It is remote control for biology.

We have known this intuitively for millennia. Every culture on earth developed some form of chanting, humming, or group singing.

Monks chant to lower their metabolic rate. Soldiers sing cadence to override fatigue and synchronize their march. Slaves sang field hollers to coordinate manual labor and survive the psychological torture of captivity. The function varied, but the mechanism was the same: sound creates stability.

Today, that stability is gone. Walk into a room of adults and ask them to sing. The reaction is visceral. They recoil. They laugh nervously. They apologize. "Oh, I can't sing," they say. "I'm tone deaf. I sound like a dying cat." They look at the floor. The shame is palpable.

This shame is a modern invention. It is a byproduct of the recording industry. Before 1877, if you wanted to hear music, you had to make it. You sang in church, you sang in the pub, you sang on the porch. There was no radio. There was no Spotify. There were no "professional" singers whose voices were pitch-corrected and compressed to perfection, beaming into your home 24/7.

When Thomas Edison invented the phonograph, he changed singing from a participatory act to a passive one. We began to compare our raw, biological voices to the polished, manufactured products of the industry. We established a standard of "good" that is mathematically impossible for the average human to meet without training and technology. We decided that if we couldn't sound like the record, we shouldn't sound at all.

We silenced the children. We told them they were "bad" at singing in music class. We cut the choir programs to fund standardized testing. We relegated singing to the shower and the car, the only places where we feel safe enough to use our own instrument. We created a culture where the only people allowed to sing are the ones who are paid to do it.

This is the equivalent of telling people they shouldn't run unless they can compete in the Olympics. We understand that running is good for the cardiovascular system, regardless of whether you are fast. We understand that painting is good for cognitive function, even if you aren't Picasso. But with singing, we accepted the lie that it is only for the elite.

The cost of this silence is high. We have lost our primary tool for social cohesion. Group singing releases oxytocin, the bonding hormone, in massive quantities. It blurs the distinction between self and other. When you sing in a choir or at a concert, your heart rate literally synchronizes with the people around you. Your breathing patterns align. You become part of a larger organism. This dissolves the ego. It fosters trust. It reduces feelings of isolation and loneliness.

In a fractured society, where we are increasingly isolated behind screens and divided by ideology, we have stopped doing the one thing that physically forces us to align. We argue, we type, we shout, but we do not sing. We have lost the ability to harmonize, both musically and socially.

This book is an investigation into what we lost and how to get it back. It is not a vocal technique manual. I do not care if you breathe from your diaphragm or lift your soft palate. I do not care if you can read music. This book is about the anatomy of the throat, the chemistry of the brain, and the anthropology of the tribe.

We will look at the science of the dropped larynx and the evolution of the rib cage. We will examine the relationship between testosterone and the deep voice, and why rock stars trigger the same biological response as tribal chieftains. We will analyze the neurochemistry of the "dopamine hit" that occurs when you belt out a chorus. We will look at stroke victims who have lost the ability to speak but can still sing perfectly, revealing that music is the brain's backup hard drive for communication.

We will travel to the delivery room, where women use low-frequency vocalization to manage the agony of labor. We will go to the funeral home, where the ancient tradition of keening uses wailing to purge grief from the body. We will look at the military boot camp, where the "Jody call" turns a group of frightened individuals into a single lethal unit.

We will debunk the talent myth. We will dismantle the idea that you are either born with a voice or you aren't. We will look at the phenomenon of "amusia"—true tone deafness—and realize that it affects less than 4% of the population. If you can tell the difference between a statement and a question, your brain can process pitch. You are not tone deaf; you are just unpracticed and ashamed.

The goal is not to get you on stage. The goal is to get you to hum. The goal is to understand that your voice is a piece of machinery that you are refusing to operate. It is a pharmacy in your throat that dispenses pain relief and antidepressants. It is a distress beacon. It is a weapon.

There is a reason the first thing we do when we enter this world is scream. We fill our lungs with air and we announce our existence. It is the most primal, honest act a human can perform. Somewhere along the way, we learned to be quiet. We learned to be polite. We learned to be consumers.

It is time to unlearn that. It is time to reclaim the noise. The silence is not serving you. It is time to open your mouth and let the glitch run.

Chapter 1

The Dropped Larynx (Anatomy of a Risk)

Take a moment to swallow.

Do it consciously. Feel the hard lump of cartilage in your throat rise up, tuck itself under the base of your tongue, and then slide back down to its resting position. That motion, which you perform roughly six hundred times a day without thinking, is the only thing standing between you and a violent death by asphyxiation. You are operating a piece of machinery that is, by any safety standard, a catastrophic design failure.

In every other mammal on earth, the tube that carries air to the lungs and the tube that carries food to the stomach are functionally separated. If you look at the throat of a dog, a cat, or a chimpanzee, the larynx—the voice box—sits high in the neck. It acts like a snorkel. The larynx locks into the back of the nasal cavity, creating a sealed airway. A dog can drink a bowl of water and breathe through its nose at the exact same time. The liquid flows around the larynx and down the esophagus, while the air flows through the larynx and into the lungs. The two streams never cross. It is a perfect, safe system.

Humans broke this system.

At some point in our evolutionary history, the human larynx detached from that safe, high position and slid down the throat. In an adult human, the larynx sits low in the neck, specifically at the level of the fourth to sixth cervical vertebrae. Because it dropped, it no longer locks into the nasal cavity. It floats free. This created a massive cavern of empty space above the vocal cords and behind the tongue, known as the pharynx.

This new gap meant that the airway and the foodway now shared a common intersection. Every time you eat a sandwich, that bolus of chewed food has to pass directly over the opening of your lungs.

To prevent the food from falling in and killing you, the larynx has to execute a complex muscular heave. It pulls up and forward, and a leaf-shaped flap of cartilage called the epiglottis folds down to cover the windpipe. It works like a toilet lid. If you laugh while eating, or if your coordination is off by a fraction of a second, the seal fails. The food drops into the trachea, the airway is blocked, and you die.

This is the "choke hazard." Humans are the only animals that routinely choke to death on their own dinner. It is a statistically significant cause of mortality. From an evolutionary perspective, this makes zero sense. Natural selection is usually ruthless about eliminating traits that cause an organism to kill itself during routine maintenance. If a mutation made a gazelle trip over its own legs every time it ran, that mutation would be wiped out in one generation. Yet, the dropped larynx persisted. It didn't just persist; it became the standard blueprint for the entire species.

The question is why nature accepted this risk. The answer is volume and variety.

The descent of the larynx converted the human throat from a simple tube into a dual-chambered resonator. That empty space in the pharynx, combined with the oral cavity, gave us two separate chambers to manipulate. By moving the tongue—which is not just the flat muscle you see, but a massive root that extends down into the throat—we can change the shape of these two chambers independently. This allows us to filter the raw sound from the vocal cords into specific vowels. We can make the distinct sounds of "ee," "ah," and "oo" because we have that vertical space to resonate them.

A chimp cannot do this. A chimp can scream, grunt, and pant, but it cannot shape the sound. Its vocal tract is a single horizontal tube. It is structurally limited to a very specific set of noises. The human vocal tract is an L-shape, with a vertical tube (the throat) meeting a horizontal tube (the mouth). This geometry allows for the articulation of complex speech and, more importantly for us, the projection of singing.

Nature decided that the ability to produce a wide range of resonant frequencies was worth the risk of occasional asphyxiation. We are built to broadcast.

To understand the broadcast, you have to look at the source signal. You have to look at the vocal folds themselves. Most people call them "vocal cords," imagining them like the strings of a guitar or a violin. This is a bad analogy. Guitar strings are passive. You pluck them, and they vibrate until the energy dissipates. If you put a camera down a human throat, you won't see strings. You will see a valve.

The vocal folds are two horizontal curtains of tissue sitting at the top of the trachea. They are white, pearly, and coated in a thin layer of mucus. They form a V-shape. When you breathe, the V opens, allowing air to pass through. When you hold your breath, the V closes, sealing the airway. This is their primary biological function: to be a sphincter. They are there to stop things from getting into the lungs and to trap air inside the lungs so you can lift heavy objects. When you grunt to lift a heavy box, you are slamming your vocal folds shut to create thoracic pressure.

Singing is the act of hacking this valve.

When you decide to make a sound, the brain sends a signal to the laryngeal muscles to bring the folds together. They don't clamp shut tight; they just touch lightly. Then, the diaphragm pushes air up from the lungs. The air hits the closed folds and pressure builds up beneath them. This is subglottic pressure. When the pressure becomes high enough, it blows the folds apart. A puff of air escapes.

As the air rushes through the gap, the speed of the airflow increases. According to Bernoulli's principle—the same physics that keeps airplanes in the sky—as the velocity of a fluid (or gas) increases, its pressure decreases. This drop in pressure sucks the vocal folds back together. They snap shut. The subglottic pressure builds up again, blows them apart again, and the cycle repeats.

This happens fast. If you sing the note A4 (the A above middle C), this cycle of opening and closing happens 440 times every single second. The tissue is colliding, blowing apart, and colliding again at a blur. It is a violent, ballistic event.

The tissue itself is a marvel of engineering. It is not a single slab of meat. It is a layered structure, specifically designed to vibrate. The deepest layer is the vocalis muscle, which provides the body and stiffness. Covering that is the lamina propria, a layer of jelly-like connective tissue. And covering that is the epithelium, a thin skin. Because the outer layers are loose and the inner layer is stiff, the vocal folds vibrate in a wave-like motion. The bottom opens first, the wave rolls up the tissue, and the top opens last. It looks like water rolling over a weir. This "mucosal wave" is what creates the complex harmonics of the human voice.

If the folds were just rigid muscle, we would sound like robotic buzzers. The gelatinous layer adds the richness, the buzz, the "human" quality to the sound. But this tissue is fragile. It requires constant hydration. It requires blood flow. If you scream too hard, you can bruise it. You can rupture a capillary. You can develop calluses (nodules) where the tissue bangs together too aggressively. We are playing a wind instrument made of wet meat.

The dropped larynx and the complex folds are the hardware, but they are useless without a power supply. You cannot run an amplifier without electricity, and you cannot run a voice without air. This brings us to the second major evolutionary shift: bipedalism.

When our ancestors stood up on two legs, they radically altered the mechanics of breathing. Consider a quadruped, like a cheetah. When a cheetah runs, its front legs hit the ground, and its internal organs—the liver, the stomach, the intestines—slam forward into the diaphragm. This forces air out of the lungs. When the cheetah pushes off, the organs slide back, pulling air into the lungs. The breath is mechanically coupled to the gait. One stride, one breath. A cheetah cannot choose to hold its breath while running. It cannot choose to exhale slowly over ten strides. It is a slave to the rhythm of its own legs.

Humans decoupled the breath from the gait. Because we walk upright, our gravity acts vertically on our organs, not horizontally against the diaphragm. Our rib cage is free. We can run and hold our breath. We can walk and exhale slowly. We can dissociate our breathing pattern from our movement pattern.

This dissociation is the battery of the voice. It allowed us to develop the "bellows" system required for singing.

The rib cage is not a rigid cage; it is a flexible basket. The ribs are attached to the spine with joints that allow them to swing up and out, like the handle of a bucket. Between the ribs are the intercostal muscles. When you inhale for a song, the external intercostals contract, lifting the ribs and expanding the chest cavity. Simultaneously, the diaphragm—a parachute-shaped muscle at the bottom of the rib cage—contracts and flattens, pushing the abdominal organs down and out. This creates a massive vacuum, and air rushes in to fill the lungs.

But the magic happens on the exhale. If you just relaxed all your muscles, the elastic recoil of the lungs would force the air out in one quick whoosh. You would make a sound for two seconds and then run out of air. To sing, you have to resist that collapse. You have to keep the rib cage expanded while the abdominal muscles slowly, precisely push the diaphragm up. You are pressurizing the tank.

This requires a level of fine motor control that is startling. You are using the large muscles of the torso—muscles designed to stabilize the spine and lift heavy weights—to dispense air in milliliters. You are maintaining a constant subglottic pressure against the vocal folds regardless of how much air is left in the lungs. Whether your lungs are 90% full or 10% full, the pressure at the throat must remain steady to keep the pitch steady.

This is why singing is physically exhausting. It is a core workout. You are turning your torso into a pressurized air column. You are overriding the body's natural desire to breathe in a shallow, rhythmic way (homeostasis) and forcing it into a pattern of rapid inhalation and extremely long, controlled exhalation.

This specific combination of anatomical accidents—the dropped larynx, the complex vocal folds, and the upright rib cage—created an instrument of terrifying power. A trained opera singer can produce sound levels of 120 decibels at a distance of one meter. That is the volume of a jackhammer. They can do this without a microphone, over a 100-piece orchestra, filling a hall that seats 3,000 people. They are doing this using nothing but the air inside their body and two strips of mucous membrane.

The efficiency of this system is high, but the vulnerability remains. The dropped larynx leaves the vocal folds exposed to the elements. Every time you breathe in, you are pulling the outside world directly onto this delicate tissue. Dust, pollen, smoke, viruses. The vocal folds are the first line of defense. They are covered in receptors that trigger the cough reflex. If a crumb hits them, they spasm. If cold air hits them, they dry out.

And yet, the biological imperative to use them is stronger than the drive to protect them. We see this in the phenomenon of the "voice change."

In biological males, during puberty, the larynx grows significantly. The thyroid cartilage—the shield protecting the front of the larynx—pushes forward, creating the Adam's apple. The vocal folds lengthen and thicken. This drops the pitch of the voice by an octave. From a survival standpoint, this is expensive. A larger larynx is more exposed to trauma. A deeper voice requires more breath energy to project.

But the signal it sends is invaluable. A low voice indicates a large larynx. A large larynx indicates high testosterone. High testosterone implies size and aggression. The deep voice is a method of peacocking without feathers. It allows a male to project size and dominance across a distance, in the dark, or through dense vegetation. It is an acoustic advertisement of fitness.

Females also undergo laryngeal changes, though less drastic. The female voice drops only a few tones, retaining the higher frequency range. This range is better suited for cutting through ambient noise—the sound of wind, rain, or other animals. High frequencies are directional; low frequencies are omnidirectional. A high-pitched scream pierces. A low-pitched roar rumbles. The combination of the two within a tribe covers the entire acoustic spectrum.

The anatomy tells the story. We are not designed for silence. If silence were the goal, the larynx would have stayed high, the ribs would be rigid, and the brain would not dedicate a massive amount of motor cortex to the lips and tongue. The cost of this apparatus is too high to be vestigial.

Consider the energy consumption. The brain is 2% of the body's weight but consumes 20% of its energy. A significant portion of that energy is dedicated to processing sound and controlling the vocal mechanism. The auditory nerve contains 30,000 fibers. The feedback loop between the ear and the voice is one of the fastest reflexes in the human body. When you hear a pitch, your brain calculates the tension required in the vocal folds to match it, sends the signal, and adjusts the muscles in milliseconds. This is the "audio-vocal loop."

When this loop is engaged, the entire body is involved. To produce a clear, resonant tone, the neck must be free of tension to allow the larynx to float. The jaw must be unhinged. The tongue must be forward. The spine must be aligned. The pelvic floor must be engaged. Singing is a whole-body event. It is not something that happens "in the throat." The throat is just the nozzle; the body is the engine.

This is why the "dropped larynx" is more than just anatomical trivia. It represents a fundamental shift in the human operating system. We moved from a species that survives by hiding to a species that survives by broadcasting. We became an open system.

In the wild, a closed system is safer. A tortoise is a closed system. It has a shell. It hides. It survives for 150 years by doing very little and making no noise. Humans are soft, shell-less, and slow. We cannot outrun a lion. We cannot overpower a chimp. Our claws are pathetic fingernails. Our teeth are flat. By all rights, we should have been eaten into extinction.

But we have the howl.

The howl allows us to coordinate. It allows us to signal alarm to the entire tribe instantly. It allows us to intimidate predators by mimicking the sound of a much larger animal. When a group of humans screams together, the volume and complexity of the sound suggest a single, massive entity.

The dropped larynx gave us the ability to fake size. By lowering the voice box, we extended the length of the tube. In acoustics, a longer tube produces a lower pitch. By rounding the lips and lowering the jaw, we extend the tube even further. We can sound bigger than we are. This is acoustic exaggeration. It is a bluff. And in the evolutionary poker game, that bluff kept us alive.

But the mechanism is fragile. It degrades. If you don't use the full range of motion of the larynx, the muscles atrophy. The cartilage calcifies as we age, turning from flexible gristle into hard bone. The vocal folds lose their elasticity. The voice becomes thin, tremulous, and weak. This is known as presbyphonia—the aging voice.

However, studies show that people who sing regularly delay this process. By constantly running the machinery through its full range of motion—stretching the folds for high notes, thickening them for low notes, expanding the ribs for deep breaths—singers keep the mechanism pliable. They maintain the collagen in the tissues. They keep the cartilage flexible. Use it or calcify.

This brings us back to the risk. The body is ruthless about efficiency. If you stop walking, your leg muscles atrophy. If you stop singing, your vocal mechanism atrophies. The brain starts to prune the neural pathways that control fine pitch. You lose the coordination between the ear and the throat. You literally forget how to sound like yourself.

The modern human throat is often a site of chronic tension. We hold stress in the jaw. We hold stress in the neck. We take shallow breaths. We sit hunched over desks, collapsing the rib cage and inhibiting the diaphragm. We are physically clamping down on the instrument. We are reversing the evolutionary advantage. We are trying to turn the larynx back into a simple valve, using it only for short bursts of speech or swallowing, and denying its function as a resonator.

When you look at the anatomy, the conclusion is unavoidable. You are built around a central column of air. You are constructed to take the atmosphere, pressurize it, and vibrate it into the world. The architecture of your skeleton, the layout of your organs, and the wiring of your nervous system all converge on this single point of exit.

The throat is a bottleneck. It is the narrowest point of the hourglass. Everything from the brain (thought, emotion, intention) has to pass through this two-centimeter gap to get to the body, and everything from the body (breath, impulse, sensation) has to pass through it to get to the world. It is a high-traffic zone.

When we block this zone—when we stop singing, when we stop shouting, when we speak in a monotone mumble—we create a traffic jam. The energy has nowhere to go. The breath gets stuck. The emotion gets stuck. The machinery jams.

The dropped larynx was a risk. Nature bet that the power of the human voice would outweigh the danger of choking. Every time you open your mouth to sing, you are validating that bet. You are proving that the risk was worth it. You are using the equipment as it was intended.

But when you stay silent, you are just a creature with a weird throat who is more likely to die eating a pretzel. You are carrying the liability of the design without reaping the benefits.

The anatomy demands action. The cartilage creates the potential. The muscles provide the power. The space creates the resonance. All of it is waiting for the signal. The instrument is sitting there, fully assembled, tuned by a million years of selection, waiting for you to simply blow air through it.

It is dangerous to have a throat like this. It is vulnerable. It is soft. But it is the only reason we are here. We did not survive because we were strong. We survived because we were loud.

Chapter 2

Peacocking with Noise (Sexual Selection)

Charles Darwin had a problem.

By the time he sat down to write The Descent of Man in 1871, he had largely figured out how natural selection worked regarding survival. He understood why a finch needed a specific beak to crack a specific nut. He understood why a polar bear needed white fur to vanish into the snow. These were practical adaptations. They solved the immediate problems of hunger and predation. But music stumped him.

He looked at the human capacity for melody and rhythm and found it biologically expensive. It burned calories. It took time away from foraging. It did not build shelter. It did not kill game. In the ruthless accounting of survival, music looked like a bad investment. Yet, it was universal. Every tribe he studied, from the Tierra del Fuego to the British aristocracy, spent a significant amount of energy making useless noise.

Darwin concluded that if the noise didn't help us survive the day, it must help us survive the generation. He wrote that the earliest human ancestors likely used their voices "in charming the opposite sex." He theorized that before we ever spoke a sentence, we sang a courtship song.

He was right. The voice is not just a tool for communication; it is a tool for seduction. It is a biological resume broadcast over the airwaves. When you open your mouth to sing, you are not just making art. You are publicly displaying the quality of your genes.

To understand this, you have to look at the animal kingdom's most famous show-off: the peacock. The male peacock drags around a massive, iridescent tail that is heavy, cumbersome, and biologically stupid. It makes him slow. It makes him easy to spot by tigers. It requires a massive amount of metabolic energy to grow and maintain.

This is the point. The tail is a signal of "honest fitness." If a bird can drag that thing around, avoid predators, and keep the feathers parasite-free and glowing, he must be genetically superior. He has energy to burn. He is telling the female, "I am so fit that I can handicap myself with this absurdity and still survive." This is the Handicap Principle, proposed by biologist Amotz Zahavi. A signal is only reliable if it is costly.

The human voice is our tail.

Singing is loud. In the Pleistocene era, being loud was a death wish. The savannah was populated by hyenas, lions, and saber-toothed cats, all of whom hunted by sound. A human humming a tune while walking to the water hole was effectively ringing a dinner bell. It signaled location, direction, and lack of vigilance.

Therefore, the act of singing in the open is a dominance display. It signals to a potential mate that the singer is not afraid. It suggests that the singer is fast enough, strong enough, or protected enough to risk revealing their position. A frightened, weak animal stays quiet. A strong, confident animal makes noise.

But the signal goes deeper than just bravery. The specific acoustic qualities of the singing voice provide a detailed medical scan of the singer's health. When a female listens to a male sing (or vice versa), she is not analyzing the lyrics. Her brain is processing the audio data to assess the singer's cardiovascular system, neurological stability, and hormonal balance.

Consider breath control. To hold a long, steady phrase, a singer must have robust lung capacity and a strong heart. The intercostal muscles must hold the rib cage open, and the abdominal muscles must regulate the pressure. If the singer has a respiratory infection, asthma, or weak lungs, the phrase will be chopped up. The breath will be audible and gasping. The tone will waver.

A smooth, sustained note is proof of cardiovascular efficiency. It tells the listener that the singer's oxygen exchange is optimal. In a hunter-gatherer society, lung power equals running power. It equals stamina in the chase. By demonstrating long phrasing, the singer is proving they can run down an antelope.

Then there is pitch accuracy. Hitting a specific note requires a high degree of fine motor control. The brain must send a precise electrical impulse to the laryngeal muscles to tension the vocal folds to an exact millimeter. If the neurological system is compromised—by age, disease, or toxin—the pitch drifts. The voice cracks.

A clear, on-pitch voice is a signal of a healthy nervous system. It shows that the neural pathways between the auditory cortex and the motor cortex are intact and firing rapidly. It is a cognitive fitness test. If you can improvise a melody, you are demonstrating processing speed and creativity, both of which correlate with intelligence. Intelligence is a highly heritable trait that increases offspring survival rates.

Tone quality acts as a flag for the immune system. The vocal folds are covered in a mucous membrane. If the body is fighting an infection, that membrane swells. The mucus becomes thick. The voice gets husky, raspy, or breathy. We can hear a cold in someone's voice before they sneeze. We can hear exhaustion. We can hear dehydration.

A "clear like a bell" voice indicates that the singer is free of parasites, free of infection, and fully hydrated. It is a certificate of health. Symmetry also plays a role. Just as we find symmetrical faces more attractive, we find "symmetrical" sounds more pleasing. If one vocal fold is slightly heavier or swollen than the other, the vibration becomes irregular. The sound becomes "rough" or "noisy" in acoustic terms. A pure tone requires perfect physical symmetry of the tissue.

Hormones are the final layer of the data stream. We are sexually dimorphic animals. Men and women sound different because of testosterone.

During puberty, testosterone floods the male system. It causes the larynx to grow and tilt, lengthening the vocal folds by about 60 percent.

It thickens the cartilage. This drops the fundamental frequency of the male voice to an average of 120 Hz.

For a female choosing a mate, a deep voice is a direct sensor for testosterone levels. Generally, the deeper the voice, the higher the testosterone exposure during development. High testosterone correlates with muscle mass, aggression, and libido. However, the data gets complicated here. Studies show that while women prefer lower voices in men, they don't necessarily want the lowest voice, which can signal excessive aggression or a compromised immune system (since high testosterone can be immunosuppressive). They want a rich, resonant baritone—the sound of balanced masculinity.

For men listening to women, the preference is usually for higher fundamental frequencies. A higher voice signals a smaller larynx and shorter vocal folds, which correlates with higher estrogen levels and youth. Since human fertility is strictly linked to age, a "youthful" voice is a primary marker for reproductive viability. The "breathy" quality sometimes associated with sexy female vocals (think Marilyn Monroe) is actually a signal of a specific laryngeal posture where the folds don't close completely, a trait more common in younger women.

This biological appraisal happens in milliseconds. You meet someone at a bar. They say "Hello" over the noise of the crowd. Before you have processed the word, your limbic system has analyzed the pitch, the timbre, and the volume. You have already decided, on a primal level, if they are a viable candidate.

If they sing, that data stream widens. Singing provides a prolonged sample. It is a stress test. Anyone can fake a "confident hello." It is much harder to fake a three-minute ballad that requires range, power, and vulnerability.

This explains the "Lead Singer Effect." In almost every modern musical group, the lead singer receives the lion's share of sexual attention. This is often dismissed as a cultural fixation on fame, but the biology suggests otherwise. The lead singer is the one taking the risk.

The drummer is sitting down. The bassist is usually standing in the back. The lead singer is at the front of the stage, often moving aggressively, projecting sound at maximum volume. They are holding the attention of the tribe (the audience). They are demonstrating stamina (running around the stage while controlling breath), cognitive skill (remembering lyrics and melodies), and social dominance (controlling the room).

To the reptilian brain, the lead singer is the alpha of the pack. They are the one howling the loudest. In a Paleolithic context, the individual who stood on the high rock and screamed at the darkness while the others watched was the protector. They were the one with the excess energy to burn. We are attracted to that energy because it promises safety and genetic potential.

The stage is just a technological amplification of the high rock. The microphone is just a prosthetic to make the signal reach more females. The leather pants are just plumage. The behavior hasn't changed in 200,000 years.

But sexual selection isn't just about showing off. It's about compatibility. This brings us to the duet.

Many bird species engage in duetting. A male and a female will sing an intricate, alternating pattern. One sings a note, the other sings the next, back and forth, so rapidly that it sounds like a single bird. This is not just a performance; it is a negotiation.

To pull off a seamless duet, the two birds must be perfectly synchronized. They must anticipate each other's rhythm. They must listen as much as they broadcast. If they can sync their songs, it signals that their neurological processing speeds are compatible. It signals that they can cooperate. Raising offspring is a cooperative act. If you can't sing together, you probably can't build a nest together.

Humans do this too. We may not chirp in rapid-fire succession, but we use music to test compatibility. We dance. Dancing is just physical singing. It is rhythmic entrainment with a partner. If you try to dance with someone and you are constantly stepping on their toes, or if they are moving to a beat that you can't hear, the attraction often dies. The "chemistry" is off. That chemistry is rhythm.

When couples sing together—whether it's in a car, a church, or a karaoke bar—they are engaging in a pair-bonding ritual. They are matching their frequencies. Harmonizing requires you to adjust your pitch relative to the other person. You have to modulate your volume so you don't drown them out. It is a microcosm of a relationship. It requires listening, adjustment, and support.

The tragedy of modern romance is that we have removed this ritual. We go to dinner. We watch movies in silence. We text. We rarely make noise together. We have replaced the active demonstration of fitness with static resumes—dating profiles that list height and income but tell us nothing about the rhythm of the person's soul.

This might explain why modern dating feels so sterile. We are trying to assess mates based on spreadsheets rather than signals. We are reading the specs of the car instead of listening to the engine.

The loss of the "serenade" is a loss of data. In pre-industrial societies, courtship often involved a significant amount of musical effort. A suitor had to learn a song, or write one. They had to perform it. This showed dedication, patience, and the ability to learn a complex skill. It was a "costly signal."

Today, sending a Spotify link costs nothing. It requires zero calories. It requires zero risk. It tells the recipient nothing about the sender other than their taste in consumption. It is a low-quality signal. Because it is cheap, it is easily ignored.

The "peacocking" aspect of singing also explains why we are so terrified of doing it. If singing is a display of genetic fitness, then singing badly is a display of genetic failure.

When you crack on a high note in public, you feel a flush of shame that is disproportionate to the error. You didn't hurt anyone. You didn't break a law. You just vibrated some air at the wrong frequency. But your brain interprets the mistake as a catastrophic loss of status. You just announced to the tribe, and to all potential mates, that your motor control is flawed, your confidence is unearned, and your equipment is faulty.

We fear singing out of tune because we fear being biologically rejected. We don't want to be the peacock with the bald patches on its tail. We don't want to be the gazelle that trips.

This fear is what keeps most people silent. We would rather hide our "tail" than risk revealing that it is less than perfect. We opt out of the game entirely. We leave the singing to the "professionals," the genetic elites who have been pre-selected for their acoustic perfection.

But this is a misunderstanding of the signal. The peacock doesn't need a perfect tail to get a mate; it just needs a tail. The signal works because it exists. The sheer act of participation is the first hurdle of fitness.

In traditional cultures, everyone sings. The village sings together. The criteria for participation is not excellence; it is presence. By opting out, you are not saving face; you are rendering yourself invisible. A silent bird gets no mate. A silent bird defends no territory.

The "bad" singer who belts out a song at a party is actually demonstrating higher fitness than the "good" singer who sits in the corner, too afraid to open their mouth. The bad singer is showing confidence. They are showing high cortisol tolerance (the ability to handle stress). They are showing social fearlessness.

Evolution rewards the bold. It rewards the noise-makers. The perfection of the note matters less than the projection of the note.

We see this in the animal world constantly. The loudest frog gets the pond. The loudest lion holds the pride. It is a contest of energy. The ability to generate sound is a proxy for the ability to generate life.

When you look at it this way, the "sex, drugs, and rock 'n' roll" cliché makes perfect sense. It is just a modern localized version of the biological imperative. The excess. The volume. The sweat. It is all a display of surplus.

We are drawn to people who sing because they are alive in a way that silent people are not. They are vibrating. They are pumping oxygen. They are broadcasting. In a world that is increasingly quiet, increasingly careful, and increasingly digital, the raw, analog signal of a human voice is a powerful aphrodisiac.

It wakes us up. It reminds us that we are animals. It reminds us that beneath the clothes and the careers and the polite conversation, we are just wet, biological machines looking for a signal in the dark.

We sing to say: "I am here. I am healthy. I am not afraid."

And the person across the room, listening, feels their own biology respond. Their heart rate syncs up. Their pupils dilate. They lean in.

Darwin was right. We didn't learn to sing to build cities. We learned to sing to build families. The city came later. The song came first.

Chapter 3

The Umbilical Chord (The First Language)

The human infant is a biological catastrophe.

Compare a newborn human to a newborn foal. A foal drops out of the womb, hits the ground, shakes off the amniotic fluid, and stands up within an hour. By the end of the first day, it can run. It is hard-wired, mobile, and functionally independent regarding its basic motor skills. A human infant, by contrast, is a helpless sack of unmyelinated nerves. It cannot lift its head. It cannot roll over. It cannot focus its eyes past eight inches. If you leave it alone on a flat surface, it will remain there until it dies of exposure or dehydration.

This helplessness is the price of our big brains. If a human fetus stayed in the womb until its brain was fully developed, the head would be too large to pass through the pelvic canal. The mother would die in childbirth, taking the infant with her. So, nature forces the eviction early. We are born effectively halfway through gestation. The first three months of life outside the womb are often called the "fourth trimester." The fetus is out, but it is not ready.

This presents a logistical nightmare for the mother. In the primal environment, a helpless infant is a heavy burden. A chimpanzee infant can cling to its mother's fur, allowing her to walk, climb, and forage with both hands free. A human infant cannot cling. It has a grasp reflex, but it lacks the strength to hold on. The mother must carry it. This occupies her arms. It burns energy. It makes gathering food difficult and escaping predators nearly impossible.

Eventually, she has to put the baby down. She has to dig for tubers, start a fire, or weave a shelter. But the moment she puts the infant on the ground, the physical connection is broken. The infant, sensing the separation, goes into immediate distress. Its heart rate spikes. It floods with cortisol. It inhales to scream. A screaming baby attracts leopards.

The mother needs a way to maintain the connection without physical touch. She needs a wireless tether. She needs a signal that says, "I am here, you are safe, do not scream," across a distance of ten feet.

That signal is the lullaby.

This is the origin of the "Umbilical Chord." It is the acoustic link that replaces the physical one. It is the first language of the human species, and it is not based on words. It is based on pitch, contour, and rhythm.

Scientists call this "Infant-Directed Speech," or IDS. Most people call it "Motherese." You know what it sounds like. You have likely done it without thinking. When an adult speaks to another adult, they use a low pitch, a fast tempo, and a flat tonal range. When that same adult speaks to a baby, their voice transforms. The pitch jumps up an octave. The tempo slows down by half. The vowels stretch out. The melody becomes exaggerated and singsong. "Hiiiiii, baaaaaby. Look at yooooou."

This is not a cultural affectation. It is not something we learned from television. It is a universal biological reflex. Field researchers have recorded mothers in the Amazon rainforest, the pulsing cities of Tokyo, the deserts of Namibia, and the suburbs of Ohio. In every single instance, regardless of the language spoken, the acoustic modification is identical. A Japanese mother speaking Japanese to her infant sounds melodically identical to a German mother speaking German to hers. They are using the same musical scale to solve the same biological problem.

The problem is attention and regulation.

The infant brain is bombarded by sensory data. It does not know how to filter the sound of the wind from the sound of a voice. It does not know what is important. Low-frequency sounds tend to blend into the background noise of the environment. High-frequency sounds cut through. By raising the pitch of her voice, the mother creates a signal that detaches from the ambient noise floor. It functions like a siren. It grabs the auditory cortex and locks it in.

But the pitch is only the carrier wave. The payload is the contour.

If you analyze the melodies of Motherese, they fall into very specific categories. These are not random tunes. They are functional commands. A short, sharp, rising melody (a staccato upward sweep) is an arousal signal. It says, "Wake up! Look at this!" Mothers use it when they want the baby to engage. A long, descending, falling melody is a soothing signal. It says, "Calm down. Sleep."

This is where the singing becomes medical. The infant's nervous system is not fully installed. The vagus nerve, which acts as the brake pedal for the heart, is immature. A baby cannot self-soothe. If it gets worked up, it stays worked up until it exhausts itself. It relies on the mother to act as an external regulator.

When a mother sings a lullaby—which is almost always a slow, repetitive, descending melody—she is physically altering the infant's physiology. The repetition of the rhythm entrains the baby's brainwaves. The slow tempo forces the baby's breathing to synchronize with the mother's. The descending pitch mimics the sound of a sigh, which triggers a parasympathetic response.

Studies show that when a mother sings to a premature infant in the NICU, the baby's oxygen saturation levels improve. The heart rate stabilizes. The cortisol levels in the saliva drop. The baby gains weight faster because it is burning less energy on stress. The song is not entertainment; it is an environmental control system. It creates a bubble of safety that allows the infant to divert energy from survival mode to growth mode.

This explains why "shushing" works. The "shhh" sound is white noise. It mimics the sound of blood rushing through the uterine arteries, a sound the fetus heard for nine months at a volume of about 85 decibels—loud as a vacuum cleaner. To a newborn, silence is terrifying. Silence means they have been abandoned.

Noise—specifically rhythmic, swooshing noise—means they are contained.

The "singing" interaction is also the gym where the brain learns language. Before a child can understand the word "apple," they must understand the concept of communication itself. They must learn that sound carries meaning.

Motherese strips language down to its musical components. By stretching the vowels ("Looooook at the dooooog"), the mother is highlighting the specific phonetic building blocks of her language. She is hyper-articulating. She is presenting the data in slow motion so the infant's auditory processor can catch it.

This is why we sing the ABCs. We don't speak them. The melody provides a container for the information. The rhythm creates a predictable structure. The brain loves prediction. If the brain can predict the next beat, it releases dopamine. If the pattern holds, the brain feels safe.

The interaction is a duet. It is not a monologue. The mother sings, and then she pauses. She waits. The baby kicks its legs or makes a cooing sound. The mother smiles and sings again. This is "turn-taking." It is the fundamental rule of human conversation. I speak, you speak. I sing, you respond.

When this turn-taking breaks down, the results are catastrophic. In the 1970s, Dr. Edward Tronick conducted the "Still Face Experiment." A mother sits facing her baby. She engages, smiles, and uses Motherese. The baby is happy, pointing, and cooing. Then, the mother stops. She turns her face into a blank mask. She does not sing. She does not smile. She stares.

Within seconds, the baby notices. It tries to re-engage her. It smiles. It waves. It makes sounds. When the mother doesn't respond, the baby begins to screech. It flails. It arches its back. It dissolves into a panic state. The withdrawal of the acoustic and emotional signal feels like a threat to survival. The baby perceives the silence as a rejection, and in the wild, a rejected baby is a dead baby.

This underscores the thesis: Singing is the primary bond. Language is the secondary bond.

We see this primacy in the way we handle pain. When a toddler falls and scrapes their knee, the mother does not offer a logical explanation of friction and dermatology. She picks the child up and she uses a specific vocal tone. She coos. She hums. "Ohhhh, ohhhh, it's okay." The words don't matter. She could be saying "The economy is collapsing" in that same tone, and the child would calm down. The melody is the medicine. The melody tells the amygdala (the fear center) to stand down.

This dynamic persists into adulthood, though we bury it. When we are sick, when we are heartbroken, when we are terrified, we regress. We want a voice that sounds like safety. We don't want a lecture. We want the tone.

The evolutionary timeline suggests that this "musical language" preceded spoken language by hundreds of thousands of years. Early hominids, likely Homo erectus, had to manage offspring with larger brains and longer dependency periods. They didn't have syntax. They didn't have grammar. But they had the throat. They likely developed a system of "contact calls"—melodic hums and grunts to keep track of each other in the tall grass.

"I am here." "Come here." "Danger." "Sleep."

These are the four pillars of the survival song. Everything else is just ornamentation.

The role of the father in this acoustic dynamic is distinct. While the mother's voice is often associated with soothing and feeding, the father's voice—lower, boomier, more resonant—is often associated with play and arousal. In many cultures, fathers engage in "rough and tumble" play, and their vocalizations reflect this. They use more staccato rhythms. They use wider pitch jumps.

This creates a stereo signal for the infant. The high-frequency, smooth signal of the mother maps the safe zone. The low-frequency, rhythmic signal of the father maps the play zone (and potentially the danger zone, as low growls are universal threat signals in nature). The infant learns to navigate the world by triangulation between these two acoustic poles.

But what happens when the signal is absent?

In modern society, we have introduced a competitor to the acoustic tether: the screen. We have replaced the mother's song with the iPad's beep. We put the baby in a stroller and hand them a device. The device makes noise. It sings songs. It has bright colors. But it lacks the critical component: contingency.

The iPad does not pause when the baby looks away. It does not speed up when the baby gets excited. It does not slow down when the baby gets tired. It just blasts data. It is a one-way street. The baby is not "in" the song; the baby is being hit by the song.

This disrupts the neurodevelopmental loop. The brain builds its circuitry based on cause and effect. "I make a sound, Mom makes a sound back." That loop builds the concept of agency. It teaches the child that they have an impact on the world. When the loop is severed, when the child is singing to a screen that doesn't care, that sense of agency atrophies.

Furthermore, the recorded voice lacks the harmonic richness of the live voice. A speaker is a vibrating paper cone. It compresses the sound. It cuts off the high and low frequencies. A human chest is a three-dimensional resonator. When a mother holds a baby against her chest and hums, the baby isn't just hearing the sound; they are feeling it. The conduction of bone and tissue transfers the vibration directly into the infant's skeleton. This tactile sound is a massive sensory input that a speaker cannot replicate.

We are seeing the results of this disconnection in the rise of speech delays and social anxiety in children. We are removing the training wheels of interaction. We are expecting children to learn to speak without first learning to sing.

This is akin to expecting someone to run a marathon without ever learning to walk. The melodic, rhythmic, emotional content of speech is the foundation. The grammar and vocabulary are just the walls and roof. If you don't pour the foundation, the house collapses.

This necessity of the "singing bond" explains why we have such a visceral reaction to the sound of a crying baby on an airplane. It is not just "annoying." It is biologically triggering. The cry is designed to be impossible to ignore. It hits a frequency—around 3,000 to 4,000 Hz—that penetrates the skull. It triggers the fight-or-flight response in every adult within earshot.

Nature designed it this way so that if a mother ignored her baby, the rest of the tribe would get so irritated that they would force her to deal with it, or deal with it themselves. It is a communal alarm system. When you are sitting in seat 14B and a baby screams, your blood pressure rises. You feel a spike of adrenaline. You are not angry; you are mobilized. Your body is preparing you to rescue the infant.

The fact that we now sit in metal tubes, strapped into chairs, unable to rescue the infant, creates a unique modern form of torture. But the instinct is ancient.

The "survival tether" also explains the phenomenon of the earworm—the song stuck in your head. The brain is a pattern-matching machine. It is designed to hold onto melodies. In a pre-literate society, if you forgot the song, you forgot the map to the water source. You forgot the history of your grandfather. You forgot the poisonous plant warning.

The infant brain is a sponge for these patterns. A child can learn the melody of a song after hearing it once. They can remember it for decades. We carry the songs of our childhood into our senility. When Alzheimer's strips away the names of our children and the ability to tie our shoes, the songs remain. The neural pathways for music are deeper, older, and more reinforced than the pathways for logic.

This is the "hard drive" that the mother is writing to. Every time she sings "Twinkle, Twinkle, Little Star," she is etching a groove into the wet cement of the developing mind. She is teaching the brain how to anticipate, how to resolve tension, and how to connect with another human being.

The tragedy is that many modern parents stop singing as soon as the child can talk. They assume the job is done. The "baby talk" is put away. They switch to "adult mode." Instructions. Questions. Commands.

"Put on your shoes." "Eat your dinner." "Stop hitting your brother."

The melody vanishes. The voice becomes a flat instrument of authority. The emotional tether thins out. The child, who learned to regulate their emotions through the rise and fall of the parent's song, is suddenly left to regulate on their own.

We maintain that the "singing" phase shouldn't end at toddlerhood. The neurological benefits of co-regulation through sound persist for life. But it starts here. It starts in the fourth trimester, when the human animal is nothing but a bundle of need, floating in a chaotic world, anchored to existence by nothing but the vibration of a woman's throat.

The umbilical cord is cut at birth. The vocal cord takes its place. It is invisible, it is weightless, but it is strong enough to haul a human soul from the darkness of instinct into the light of consciousness.

Chapter 4

The Pharmacy in Your Head (Neurochemistry)

If you want to change your state of mind, you usually have to ingest something. You drink coffee to wake up. You drink alcohol to calm down. You smoke a cigarette to focus. You take an aspirin to stop the pain. We are a species of chemistry. We understand that if we want to alter the way we feel, we need to introduce a foreign substance into the blood. We treat the body like a car that needs specific fluids topped up to run smooth.

This reliance on external inputs ignores the fact that you are walking around with a fully stocked pharmacy in your head. Your brain is capable of synthesizing and releasing compounds that are functionally identical to morphine, cocaine, and Valium. It does not charge a copay. It does not require a prescription. It only requires a trigger.

For the last 50,000 years, the primary trigger was the human voice.

When you open your mouth and sustain a tone, you are not just making noise. You are hacking your own endocrine system. You are manually engaging a feedback loop that forces the brain to manufacture and release a specific cocktail of neurotransmitters. This is not a metaphor. It is a measurable biological event. If you hook a singer up to an MRI or take a saliva sample before and after a choir rehearsal, the data is stark. They are high.

The first drug on the shelf is dopamine. This is the reward chemical. It is the same neurotransmitter that floods the brain when you eat sugar, when you have sex, or when you win a hand of poker. It is the currency of desire. The brain uses it to reinforce behaviors that are essential for survival. "You did a good thing," the brain says. "Here is a cookie."

In the context of singing, dopamine is tied to prediction. The human brain is a prediction machine. It is constantly guessing what is going to happen next. When you listen to a melody, your auditory cortex is anticipating the next note. It knows the scale. It knows the rhythm. It builds a model of the future.

When you are the one singing, the stakes of that prediction are higher. You are not just guessing the note; you are executing it. You have to coordinate the breath, the vocal cords, and the vowels to hit a target that exists only in your mind.

When you hit it—when the physical sound in the room matches the mental image in your head—the brain registers a "prediction error" of zero. The tension resolves. The nucleus accumbens, a small region deep in the brain associated with reward, dumps dopamine into the system. You feel a rush.

This is the "shiver" or the "chill" you get when a song hits its climax. It is a biological reward for accuracy. The brain is paying you for successfully navigating a complex motor task. It is the same chemical hit a basketball player gets when the ball swishes through the net. It is the thrill of competence.

But the dosage increases with volume and effort. Singing loud requires more energy, more risk, and more coordination than singing soft. Belting a high note is a high-risk maneuver. If you miss, you crack. If you hit it, the dopamine reward is scaled to the difficulty. This is why singing in the shower feels better than humming. You are betting more chips, so the payout is bigger.

The second drug is oxytocin. This is often called the "cuddle hormone" or the "love drug," which makes it sound soft. It is not soft. It is the chemical of tribal survival. It is the glue that keeps mammals from eating their young. It is released during breastfeeding, during orgasm, and during hug. It lowers the barrier between "self" and "other."

Singing is the only passive human activity that releases oxytocin in massive quantities. You don't have to touch anyone. You don't have to know them. You just have to breathe with them.

When a group of people sings together, their breathing patterns synchronize. This is unavoidable. You cannot sing the phrase if you don't inhale at the start of it. Within two minutes of a choir song, every person in the room is inhaling and exhaling at the exact same moment.

This synchronization tricks the brain. In the wild, the only time you would breathe in perfect sync with another human is if you were engaging in a highly coordinated physical act—mating, hunting, or running from a predator. The brain interprets this synchronization as deep intimacy. It assumes that these people are your tribe. It assumes they are safe. So, it releases oxytocin.

This explains the "instant bond" of the choir or the football terrace. You can stand in a stadium with 50,000 strangers, people you would never talk to on the street, people who vote differently than you and live in different neighborhoods. But when the anthem starts, and you all hit the same rhythm, the walls come down. You feel a sudden, intense affection for the mass of humanity around you. You trust them.

This is chemical warfare against isolation. We live in an era of extreme loneliness. We are connected digitally but separated physically. We rarely touch. We rarely breathe together. We are oxytocin-starved. Singing is the most efficient delivery system for this compound. It creates a synthetic village. It allows you to feel the safety of the herd without the politics of the herd.

The third drug is the heavy hitter: endorphins. These are endogenous opioid peptides. They are the body's natural morphine. They bind to the same receptors in the brain as heroin and oxycodone. Their job is to block pain and induce euphoria.

The body releases endorphins during extreme physical exertion or trauma. The "runner's high" is an endorphin dump. The body realizes you are pushing it to the limit—tearing muscle fibers, depleting oxygen—so it numbs the pain to keep you moving.

Singing is an athletic event. It engages the intercostal muscles, the abdominals, the diaphragm, and the microscopic muscles of the larynx. It creates high internal pressure. It vibrates the skull. To the brain stem, this looks like work. It looks like stress.

So, the brain dispenses the painkillers.

This is why singing feels good even when you are sad. It's why the "Blues" works. You are singing about misery, you are crying, you are grieving, but physically, you are stoned on your own supply. The act of singing metabolizes the pain. It turns emotional suffering into a physical event that the body can treat with opioids.

Studies on choir members show that their pain tolerance increases significantly after rehearsal. If you put your hand in a bucket of ice water, you will pull it out after a certain number of seconds. If you sing for an hour and then put your hand in the bucket, you can hold it there longer. You are chemically buffered.

This analgesic effect is immediate. It doesn't take weeks of practice. It happens the first time you really push air through the pipes. It is the reason we sing in chains. It is the reason we sing in fields. It is not just to pass the time; it is to anaesthetize the body against the labor.

The final component of the cocktail is the removal of a toxin: cortisol.

Cortisol is the stress hormone. It is designed to save your life. When you see a tiger, cortisol spikes. It shuts down digestion, halts the immune system, and dumps glucose into the blood for quick energy. It prepares you to fight or run.

In the modern world, we rarely see tigers. We see emails. We see traffic. We see bills. The problem is that the brain cannot distinguish between a tiger and a tax audit. It triggers the cortisol response regardless. Since we don't run or fight, the cortisol stays in the blood.

It marinates the organs. Chronic cortisol exposure destroys the body. It causes heart disease, weight gain, sleep disorders, and anxiety.

Singing is the most effective natural method for lowering cortisol.

The mechanism is the vagus nerve, which we touched on earlier, but the chemical result is the key here. When you stimulate the vagus nerve through the long, controlled exhalations of singing, you are sending a biochemical "all clear" signal. You are telling the adrenal glands to stand down.

Clinical studies have compared the cortisol levels of people who listen to relaxing music versus people who sing. Listening helps a little. Singing helps a lot. The active participation—the physical vibration and the breath control—is what flips the switch. You have to do the work to get the clearing effect. You cannot outsource the regulation to a pair of headphones.

This reduction in cortisol has a secondary effect: it boosts the immune system. High cortisol suppresses the production of antibodies. When you lower the cortisol, the immune system wakes up. It starts producing Immunoglobulin A, an antibody that fights infections in the respiratory tract.

Literally, singing protects you from getting a cold. By making noise, you are reducing the stress load that leaves you vulnerable to viruses.

When you combine these four elements—dopamine, oxytocin, endorphins, and cortisol reduction—you have a potent neurochemical event. You have a state of mind that is alert (dopamine), connected (oxytocin), pain-free (endorphins), and calm (low cortisol).

This specific state is hard to achieve by other means. You can get dopamine from a video game, but you don't get the oxytocin. You can get oxytocin from sex, but you don't necessarily get the group connection (unless it's a very specific kind of party). You can get endorphins from a marathon, but that takes four hours and ruins your knees.

Singing delivers the full package in about three minutes.

This "pharmacy" model explains why we get addicted to music. It explains why we hum when we are nervous. We are self-medicating. We are trying to balance the ledger.

Consider the teenager in the bedroom, screaming along to a track on the radio. The parents yell, "Turn that down!" They see it as noise. They see it as rebellion. But the teenager is managing a chaotic internal environment. Their brain is remodeling. Their hormones are rioting. They are flooded with new emotions they cannot name or control.

When they sing, they are taking the controls. They are forcing a dopamine hit to counter the depression. They are burning off the cortisol of high school social stress. They are using the voice to survive the chemistry of adolescence.

Or consider the Alzheimer's patient. The disease has eaten the hippocampus. Short-term memory is gone. They don't know what they had for breakfast. They don't know their daughter's name. The logic centers are dark.

But if you play a song from their youth, they light up. They sing every word. For three minutes, they are back. They are lucid.

Why? Because the musical memory is stored in a different part of the brain, often the pre-motor areas and the basal ganglia, which are more resistant to the atrophy of the disease. But more importantly, the act of singing triggers that chemical wash. It wakes up the remaining neural networks. It floods the dark room with light. It is the last medicine that works.

The tragedy is that we have restricted access to this pharmacy. We have decided that you need a license to use it. If you are not a "good singer," you are told to keep the cabinet locked.

Imagine if we did this with running. Imagine if we told people, "Unless you can run a five-minute mile, you shouldn't jog. It looks bad. You look silly." We would have a population dying of heart disease. We understand that the benefit of exercise is in the doing, not the performance.

We have not made that leap with singing. We still judge the output rather than the input. We critique the pitch and ignore the chemistry.

This creates a vicious cycle. People are stressed, lonely, and in pain. They have a tool to fix it—their voice. But they are too ashamed to use it. So they stay stressed, lonely, and in pain. They go to the doctor. They get a prescription for Xanax. They get a prescription for opioids. They seek the chemical fix from the outside because they have been shamed out of generating it on the inside.

The pharmaceutical industry is a trillion-dollar business. The vocal industry is a talent show. We are missing the point. The value of the human voice is not in the record sales. It is in the public health.

This is not to say that singing cures cancer or fixes a broken leg. It is not magic. It is maintenance. It is the daily hygiene of the nervous system. Just as you brush your teeth to prevent decay, you should sing to prevent decay of the spirit.

The "dose" required is not high. You don't need to join a professional choir. You don't need to perform an aria. The biological response is triggered by the vibration, not the applause. Singing in the car works. Humming while you do the dishes works.

The key is the vibration. The skull must buzz. The chest must rumble. You have to physically occupy the sound.

There is a distinct difference between the chemistry of listening and the chemistry of singing. Listening is passive. It can trigger emotion, yes. It can trigger memory. But it does not require the motor cortex. It does not engage the breath. It does not force the synchronization of the heart rate to the rhythm in the same way.

Listening to a song is like watching someone else work out. You might get inspired, but you don't get the muscle growth. To get the chemicals, you have to do the reps.

This biological reality puts the "Talent Myth" in a new light. If singing is a health mechanism, then "talent" is irrelevant. You don't have to be talented at breathing to get the benefits of oxygen. You don't have to be talented at sleeping to get the benefits of rest. You just have to do it.

The person who sings off-key gets the exact same chemical benefit as the person who sings perfectly. The dopamine doesn't care about the pitch; it cares about the prediction and the effort. The oxytocin doesn't care about the tone; it cares about the synchronization. The endorphins don't care if you sound like a dying cat; they care that you are expelling air under pressure.

In fact, the "bad" singer might get a bigger hit. They are working harder to find the notes. They are overcoming more social resistance. The exertion is higher.

We need to reframe the conversation. We need to stop talking about "voice" as an aesthetic object and start talking about it as a physiological function. We need to see the throat as a regulator.

When you feel the tightness in your chest—that familiar, modern constriction of anxiety—you have two choices. You can wait for it to pass, or you can vibrate it out. You can swallow the pill, or you can make the sound.

The pharmacy is open. It is open 24 hours a day. It is located between your chin and your collarbone. And the only way to get the medicine out is to open your mouth and push.

Chapter 5

Right Brain Rebellion (Neurology)

On January 8, 2011, a bullet entered the skull of Congresswoman Gabrielle Giffords above her left eye, passed through the left hemisphere of her brain, and exited the back of her head. The trauma was catastrophic. In a fraction of a second, the shockwave turned the delicate architecture of her neural tissue into a chaotic ruin. She survived the surgery. She survived the coma. But when she woke up, the world was silent.

The bullet had ravaged Broca's area. This is a small, walnut-sized region in the left frontal lobe responsible for speech production. It is the command center that translates abstract thought into motor commands for the mouth and tongue. Giffords had "expressive aphasia." She was fully conscious. Her intelligence was intact. She knew who she was, she knew her husband, and she knew she was in a hospital. But she could not say a word. The bridge between the mind and the mouth had been blown up.

For a human being, this is a specific kind of hell. It is "locked-in" syndrome without the paralysis. You are screaming inside a soundproof room. The thoughts are vivid, but the mechanism to export them is dead.

Then, her therapist tried something that looked, on the surface, like a parlor trick. She didn't ask Giffords to speak. She asked her to sing.

The therapist started a simple, rhythmic melody. "Happy... Birth... Day..."

Giffords, who could not say "hello," opened her mouth. She took a breath. And she sang, "To... You."

She didn't stumble. The words came out clear, on pitch, and in the correct order. The room did not witness a miracle; it witnessed a rerouting. The bullet had destroyed the left brain's speech center, but it had missed the right brain's music center. The highway was closed, so the brain took the service road.

This phenomenon is the "Right Brain Rebellion." It is the definitive proof that music is not just a decorative software running on top of our operating system. It is a fundamental, redundant hard-wired system designed to keep us communicating even when the primary circuits fail.

To understand how this works, you have to look at the real estate of the brain. Since the mid-19th century, neurologists have known that the brain is lateralized. For roughly 95% of right-handed people, language lives on the left. Broca's area handles the production of speech (finding the words and moving the mouth), and Wernicke's area, located further back, handles the comprehension of speech (understanding what you hear).

The right hemisphere was historically dismissed as the "minor" hemisphere. It was thought to be silent, emotional, and largely supportive. But the right hemisphere houses the prosodic functions of language. It handles the melody of speech—the rise and fall of pitch that indicates sarcasm, a question, or anger. It handles rhythm. It processes music.

When you speak a sentence like "I am hungry," your left brain handles the syntax (Subject-Verb-Adjective) and the vocabulary. Your right brain handles the intonation. Usually, these two sides talk to each other constantly across the corpus callosum, the thick band of nerve fibers connecting the hemispheres. They work in tandem.

But when the left side is destroyed—by a bullet, a stroke, or a tumor—the right side is left isolated. The words are gone because the left-side file cabinet is locked. But the melody is still there.

This is where Melodic Intonation Therapy (MIT) comes in. It was developed in the 1970s, but its roots go back to observations made in the 1800s by Paul Broca himself, who noted that patients who could only say the word "tan" could still sing the Marsellaise perfectly.

MIT is not singing lessons. It is a brute-force method of hacking the brain's plasticity. The therapist sits across from the patient. They take the patient's left hand and tap it on the table in a steady rhythm. Tap. Tap. Tap. This engages the right hemisphere's motor cortex (which controls the left side of the body) and its rhythm processing centers.

Then, they intone a phrase. They don't say "I want water." They sing it on two distinct pitches, usually a minor third interval, dragging out the vowels. "I... waaaant... waaa... ter."

The patient mimics the sound. At first, they are just mimicking the melody. They are bypassing the broken Broca's area entirely. They are using the right hemisphere's "homologous" area—the mirror-image region of Broca's area—to drive the mouth.

This is the backup drive. The brain has a redundant system for vocalization, but it is normally dormant for speech. It is reserved for singing. When the primary drive fails, MIT forces the secondary drive to spin up.

The results are physical. If you scan the brain of a severe aphasic patient before therapy, the left side is dark and the right side is quiet. After months of MIT, the right side lights up. The "singing center" actually grows. The grey matter density increases. The white matter tracts—the wiring—thicken. The right hemisphere literally learns to speak.

This is neuroplasticity in its most aggressive form. The brain realizes the left side is dead, so it colonizes the right side. It repurposes neurons that were designed for melody and trains them to handle vocabulary. Giffords eventually learned to speak again, not just sing. She used the singing as a scaffolding to rebuild the speech.

This redundancy suggests that nature places a premium on communication. In engineering, you only build a backup system for a critical component. You have two kidneys. You have two lungs. You have two eyes. You do not have two livers or two hearts. If the liver fails, you die. If speech fails, you are socially dead. So, evolution gave us a spare tire. It hid the spare tire inside a song.

The "Right Brain Rebellion" isn't just for stroke victims. It is also the mechanism that liberates the stutterer.

Consider the case of King George VI, famously depicted in The King's Speech. Or the singer billie Eilish. Or James Earl Jones. Or Kendrick Lamar. All of them struggled with severe stuttering. A stutter is a motor timing glitch. It is a neurological misfire in the basal ganglia, the deep brain structure that initiates movement.

When a stutterer tries to speak, the brain sends the signal to the mouth, but the timing is off. The signal gets jammed. The first syllable repeats like a broken record, or the jaw locks up in a silent block. It is a terrifying loss of control. The harder they try to force the word out, the worse the jam becomes. Anxiety spikes, cortisol floods the system, and the motor loop tightens.

But ask a stutterer to sing, and the stutter vanishes. Instantly.

This is not a gradual improvement. It is a binary switch. A man who cannot say his own name without struggling for thirty seconds can sing an intricate opera aria without a single hesitation.

Why? Because singing runs on a different clock.

Speech is "self-generated." You decide what to say, and your brain has to internally generate the timing for every syllable. It has to coordinate the lips, tongue, and larynx in a free-form jazz improvisation of millisecond precision. For a stutterer, this internal clock is faulty.

Singing is "externally generated." Even if you are singing alone, you are following a pre-existing structure. The melody dictates the rhythm. The duration of the note is fixed. The breath is planned. You are not improvising the timing; you are stepping onto a moving walkway.

When you sing, the brain shifts control from the faulty basal ganglia loop to the premotor cortex and the auditory cortex. The rhythm acts as a pacemaker. It overrides the glitch. The "right brain" provides the timeline, and the motor system simply follows the beat.

This creates a profound sense of liberation. For the stutterer, the singing voice is the only time they are fluent. It is the only time the machinery works as intended. This is why so many stutterers become professional singers. It is not a coincidence. It is a survival strategy. If you cannot talk to the tribe, you sing to them.

This duality—the split between the speaking brain and the singing brain—hints at the evolutionary timeline. Music likely came first. The neural pathways for rhythm and pitch are deeper and more robust than the pathways for syntax.

We see this in dementia. We see it in Alzheimer's. The "Pre-Historic Hard Drive" is the last thing to be wiped.

Visit a memory care unit. You will see patients who are catatonic. They sit in wheelchairs, heads down, drooling. They do not recognize their children. They have not spoken a coherent sentence in years. The left brain has been eaten away by the plaques and tangles of the disease. The vocabulary is gone. The logic is gone.

Put headphones on them. Play a song from their youth. Play a hymn.

The reaction is seismic. The head lifts. The eyes open. The foot starts to tap. And then, they sing. They sing every word. They sing the verses. They sing the chorus. They harmonize.

For three minutes, the person returns. The "Right Brain" is still online. The musical memory is stored in a distributed network that involves the cerebellum (muscle memory) and the emotional centers of the limbic system, areas that are often spared until the very late stages of the disease.

This is not just "comfort." It is reconnection. When the music stops, the window often stays open for a few minutes. The patient might turn to you and say, "That was beautiful," or "I remember that." The music jump-started the system. It flooded the engine with dopamine and woke up the dormant pathways.

This implies that our definition of "brain damage" is often too narrow. We look at the scan, see the hole in the left hemisphere, and declare the communication dead. We give up. We stop talking to the patient because they don't talk back.

But the singing brain is resilient. It is the cockroach of neurology—it survives everything. It survives trauma, it survives degeneration, it survives the bullet.

The implications for the average person—the one without a bullet hole in their head—are less dramatic but equally functional. We all have a "split brain." We all have a logical, linear left side that wants to organize the world into spreadsheets, and a holistic, rhythmic right side that wants to feel the world as a pattern.

Modern education and corporate culture are aggressively left-brained. We prize analysis. We prize data. We prize the specific word. We starve the right hemisphere. We stop singing when we leave elementary school, and we let that neural real estate lie fallow.

When we do this, we lose our capacity for "holistic" processing. The right brain is better at seeing the big picture. It is better at emotional nuance. It is better at empathy. When we stop singing, we are turning off the lights in half the house.

Using the voice integrates the hemispheres. When you sing a song with lyrics, you are forcing the left brain (words) and the right brain (melody) to cooperate. You are firing signals across the corpus callosum like a machine gun. You are synchronizing the two halves of the processor.

This is why singing helps with learning. If you want to memorize a list of facts, you are using your left brain. If you set that list to a melody, you are bringing in the right brain. You are doubling the processing power. You are creating two separate retrieval cues—one linguistic, one musical. If the left brain forgets the word, the right brain remembers the tune, and the tune drags the word back up with it.

This is the "ABC" effect. You cannot say the alphabet backward without thinking hard, but you can sing it forward without thinking at all. The melody is a container.

The neuroplasticity argument suggests that we should be using singing as a preventative medicine for the brain. Just as we do crossword puzzles to keep the left brain sharp, we should be singing to keep the right brain thick. A thicker right hemisphere is a better backup generator. If you ever do have a stroke, if you ever do suffer cognitive decline, you want that right hemisphere to be robust. You want those pathways to be superhighways, not dirt roads.

The story of Gabby Giffords is a tragedy, but it is also a manual. It shows us the wiring diagram. It shows us that the human voice is not a single instrument. It is a dual-interface system.

When the world breaks you, when the logic fails, when the words are stolen, the melody remains. It is the last room in the house to burn down.

Nature did not design us to be silent. It designed us to be redundant. It gave us a way to speak when we cannot speak. It gave us a way to connect when the connectors are severed.

The rebellion is always waiting. It is sitting in the right temporal lobe, waiting for the downbeat. It is the survival protocol that kicks in when the logic centers crash.

We spent millions of years evolving this redundancy. To let it atrophy because we are "embarrassed" to sing is a waste of biological engineering. It is like owning a jet and refusing to fly it because you are afraid of heights.

The brain is plastic. It changes based on what you ask it to do. If you ask it to be silent, it will prune the connections. It will shrink the right hemisphere. It will dismantle the backup drive to save energy.

But if you ask it to sing, it will build. It will lay down new myelin. It will sprout new dendrites. It will keep the bridge open.

Singing is the maintenance crew for the mind. It keeps the lights on. It keeps the road clear. And sometimes, when the bullet hits, it is the only thing that gets you home.

Chapter 6

The Pre-Historic Hard Drive (Memory & History)

Try to recite your credit card number.

Do it right now. Close your eyes and say the sixteen digits in order. Unless you are a savant or you use that card twenty times a day, you probably stalled after the second block of four. The brain struggles to hold a random string of numbers. It is data without a container. It slips through the cognitive mesh.

Now, sing the theme song to The Fresh Prince of Bel-Air.

You likely didn't even have to think about it. The lyrics unspooled instantly. "In West Philadelphia born and raised..." You know the rhythm. You know the pitch. You know exactly where the pauses are. You haven't heard that song in ten years, yet the file is sitting on your desktop, fully corrupted-free, ready to play.

This discrepancy is not an accident. It is a feature of the human operating system known as the "mnemonic hack." For 99% of human history, we did not have writing. We did not have ink, paper, or hard drives. We had the air, and we had the ear. If you knew something—how to find water, which berries killed you, the genealogy of your tribe—and you didn't pass it on, that information died with you.

The problem is that the human brain is terrible at rote memorization of dry facts. It dumps non-essential data constantly to save energy. To survive, our ancestors needed a way to lock information into the neural circuitry so that it could not be deleted. They needed a compression format.

They found it in melody.

Melody is a data structure. When you attach a piece of information to a tune, you are not just making it pretty; you are giving it a skeleton.

You are imposing a strict set of rules—rhythm, rhyme, and pitch—that limits the possibilities of what the next word can be.

If you are trying to remember a sentence in prose, the next word could be anything. If you are trying to remember a line in a song, the next word has to fit the meter. It has to fit the rhyme scheme. It has to match the melodic contour. The constraints act as error-correction code. If you sing the wrong word, it physically feels wrong in your mouth because it breaks the rhythm. The song self-corrects.

This is why the Torah is chanted, not read. This is why the Quran is recited with specific melodic cadences (Tajweed). This is why the Vedas were passed down for three thousand years without a single syllable changing, long before they were written down. The melody locked the data in place.

We tend to think of history as something found in books. But for most of our existence, history was a song.

Consider the Iliad and the Odyssey. We read them today as thick paperback novels in translation. We study them as literature. But they were never intended to be read. They were performance pieces. They were audiobooks broadcast live.

Scholars Milman Parry and Albert Lord revolutionized our understanding of this in the 1930s. They traveled to the Balkans to record illiterate Serbian bards who were still singing epic poetry. These men could perform songs that were thousands of lines long. They could sing for hours without stopping.

Parry and Lord discovered that these singers weren't memorizing the text word-for-word like an actor memorizing a script. They were using a system of "formulas." They had a library of stock phrases that fit specific rhythmic slots in the meter (dactylic hexameter). "Swift-footed Achilles." "Wine-dark sea." "Rosy-fingered dawn."

These weren't just poetic flourishes. They were modular building blocks. If the singer needed to fill the second half of a line, he grabbed a pre-fabricated block that fit the rhythm. The melody and the meter provided the grid, and the singer dropped the data into the slots. This allowed them to "write" the story in real-time, accessing a massive database of cultural history without needing a single scrap of paper.

The music was the retrieval system. Without the rhythm, the bard would stumble. With the rhythm, he could access the archives.

This function of song as a "survival map" is most clearly seen in the Aboriginal cultures of Australia. For tens of thousands of years, they have navigated a continent that is largely desert using "Songlines."

A Songline is a track across the land. It is a physical path, but the map exists entirely in audio. The song describes the land. It describes the specific rock formations, the waterholes, the trees, and the turns in the creek bed. As the traveler walks, they sing the song. The rhythm of the song corresponds to the walking pace.

If the song speeds up, the walker knows they need to move faster to reach the next water source before nightfall. If the melody dips low, it might indicate a valley. If the lyrics describe a "Lizard Man" hiding behind a rock, it marks a specific geographical landmark that looks like a lizard.

This is not a metaphor. It is a GPS system. A person who knows the song can navigate territory they have never seen before, simply by following the instructions encoded in the melody. If they stick to the song, they live. If they forget the song, they die of thirst in the bush.

The song also encodes the law. It tells you which animals you can eat and when. It tells you who you can marry to avoid inbreeding. It tells you the boundaries of your territory. The entire legal and ethical code of the civilization is stored in the music.

This explains why the destruction of culture often begins with the banning of songs. When colonial powers invaded, they didn't just take the land; they outlawed the traditional dances and chants. They understood, perhaps instinctively, that the song was the library. If you silence the song, you delete the history. You reset the hard drive.

In the modern world, we have outsourced our memory. We have Google. We have Wikipedia. We have contact lists. We don't need to remember anything. Consequently, our "memory muscles" have atrophied. We panic if we lose our phone because we have outsourced our brain to it.

But the biological mechanism is still there. We see it in the way we teach children.

You want a child to learn the alphabet. You don't hand them a list. You teach them the "Twinkle, Twinkle, Little Star" melody. You map the letters A through G to the first phrase. H through P to the second. The melody acts as the glue.

You want a medical student to remember the cranial nerves. They learn a rhyming mnemonic. You want a math student to remember the quadratic formula. There is a jingle for that.

The brain prefers this format because music is "multi-modal." When you read a text, you are primarily using the visual cortex and the language centers. It is a relatively "thin" trace.

When you sing a song, you are lighting up the whole house. You are using the motor cortex to control the vocal cords and the breath. You are using the auditory cortex to hear the pitch. You are using the limbic system to feel the emotion of the melody. You are using the cerebellum to keep the rhythm.

This creates a "thick" trace. The memory is anchored in multiple locations. It is tied to movement, sound, and feeling simultaneously. If one anchor fails, the others hold.

This robustness is why music is the last thing to go when the brain begins to rot.

Alzheimer's disease is a terrifying dismantling of the self. It typically starts in the hippocampus, the area responsible for forming new short-term memories. The patient forgets where they put their keys. Then they forget what they did yesterday. Then the damage spreads to the temporal lobes, erasing the semantic memory—the names of objects, the names of people.

Eventually, the patient is left in a perpetual present, untethered from their past. They lose the narrative of who they are.

But the musical memory is stored differently. Neuroimaging studies show that the areas of the brain that store musical memory—specifically the caudal anterior cingulate and the ventral pre-motor cortex—show significantly less atrophy in Alzheimer's patients than the rest of the brain. These areas are evolutionarily older. They are tougher.

This leads to the "Clive Wearing" paradox. Clive Wearing was a British musicologist who suffered a severe brain infection (herpes simplex encephalitis) in 1985. The virus destroyed his hippocampus completely. He has a memory span of about seven to thirty seconds. Every time he blinks, it is like he is waking up for the first time. He greets his wife with shocking joy every time she walks in the room, even if she just went to the kitchen for water, because he believes he hasn't seen her in years.

He cannot remember reading a sentence by the time he gets to the end of it. He is trapped in a terrifying, looping now.

But sit him at a piano.

He plays Bach. He plays complex, intricate fugues. He sight-reads music. He conducts a choir. While he is inside the music, he is fully functional. He knows where the piece is going. He knows the structure. He anticipates the key changes.

The moment the music stops, he crashes back into the void.

The music provides a temporal bridge. It allows him to string seconds together into a coherent minute. It is the only time his brain can organize time.

This is not rare. Visit any nursing home. You will see patients who are non-verbal, slumped in wheelchairs, eyes vacant. They are classified as "unresponsive." Play a song from their teenage years—say, "You Are My Sunshine" or a big band hit.

The transformation is mechanical. The head lifts. The eyes focus. The mouth opens. They sing the lyrics. Not just the chorus, but the verses. They access a file that was written to the disk sixty years ago, a file that has survived the corruption of the operating system.

This "reawakening" is often distressing to family members because it reveals that the person is still in there. They aren't gone; they are just inaccessible. The door is locked, but the music slides a key under the frame.

We use this practically in end-of-life care. Hospice workers know that hearing is the last sense to fade. A dying brain, shutting down organ by organ, can still process melody. Singing to the dying is a cross-cultural universal. It is the bookend to the lullaby. We sing them in, and we sing them out.

The song provides a container for the transition. It regulates the breathing of the dying person, much like it regulates the breathing of the infant. It creates a shared frequency for the people in the room, who are paralyzed by grief. It gives them something to do.

In this context, the "song" is not entertainment. It is a utility. It is a tool for transition.

The durability of musical memory also explains the persistence of trauma. We don't just remember the good songs. We remember the bad ones.

A specific song can trigger a PTSD response instantly. If a soldier heard a certain track on the radio before an IED explosion, that song is forever welded to the event. The brain encoded the melody and the trauma together in the amygdala. Hearing the song twenty years later doesn't just remind him of the event; it physically recreates the physiological state of the event. The heart races. The palms sweat. The time collapses.

This is the "Hard Drive" working too well. It has set the file to "Read Only" and "Auto-Play."

Commercial industries understand this stickiness. This is why advertising jingles exist. Advertisers know that if they say "Buy insurance," you will forget it. If they sing "Nationwide is on your side," you will carry that branding in your skull for the rest of your life against your will. They are hacking your evolutionary retention software to sell you product.

They are exploiting the fact that your brain is desperate to finish the pattern. If you hear "Ba-da-ba-ba-ba," your brain screams "I'm lovin' it." You cannot stop it. The neural circuit fires automatically. It is a hijack.

The shift from oral culture to written culture has changed us. We have gained precision, but we have lost durability. A digital file can be corrupted. A book can burn. A server can crash.

But a song that is passed from mother to child, generation after generation, is remarkably resilient. It survives genocides. It survives diaspora.

Consider the African American spirituals. When enslaved people were brought to America, they were stripped of their languages, their names, and their gods. They were forbidden to read or write. They were forbidden to own drums.

But they could sing.

They encoded their resistance, their history, and their maps in the spirituals. "Wade in the Water" wasn't just a religious hymn; it was a tactical instruction. It told runaways to get into the river so the bloodhounds would lose their scent. "Follow the Drinking Gourd" was an astronomical map pointing to the Big Dipper and the North Star.

The slave owners heard "noise" or "piety." The enslaved heard "survival data."

The song was a steganographic container—hidden writing. It allowed a captive population to communicate openly in front of their captors without being detected. It preserved a sense of community and shared destiny in a system designed to atomize individuals.

If they had written these instructions down, they would have been found and executed. By singing them, they kept the data invisible and mobile. You cannot search a person's throat for contraband.

This historical reality reinforces the idea that singing is a technology of the oppressed. It is the tool of those who have no other tools. When you have no power, no weapons, and no platform, you still have the vibration of air. You still have the ability to encode your truth in a melody and broadcast it.

In our own lives, we feel the faint echo of this prehistoric hard drive.

Think about how you learn a foreign language. The hardest part is the vocab and the grammar. But if you learn a song in that language, you learn the pronunciation perfectly. You learn the rhythm of the sentence.

Think about how you remember the days of the month. "Thirty days hath September..." You chant it. You add a cadence. You turn it into a primitive song.

We are constantly reaching for this tool, even though we live in the information age. We instinctively know that if we want to keep something, we have to make it rhythmic.

This suggests that our dismissal of "rote memorization" in schools might be a mistake. We stopped asking kids to memorize poetry. We stopped asking them to memorize dates. We told them, "You can just look it up."

But "looking it up" is not knowing. Knowing is carrying the data in your body. When you memorize a poem or a song, it becomes part of your furniture. It is there to comfort you, to guide you, to provide a structure for your thoughts when the internet is down.

The "Pre-Historic Hard Drive" is still spinning. It is waiting for data.

When we stop singing, we stop writing to the disk. We rely entirely on the fragile, external memory of our devices. We become empty vessels, drifting on a sea of information but holding none of it.

We need to start hitting "Save" again.

We need to memorize the songs of our own lives. We need to sing the stories of our parents. We need to teach our children melodies that contain our values, not just melodies that sell them burgers.

Because one day, the lights might go out. The screen might go black. The book might crumble. And when that happens, the only history that will survive is the history we can sing.

The voice is the only archive that cannot be burned. It is the only library that walks away when the city falls.

Chapter 7

Weaponized Sound (War & Intimidation)

The New Zealand All Blacks are not a choir. They are a rugby team. They are large men, built like refrigerators, designed to inflict blunt force trauma on other large men. Before every international match, they stand in a formation on the pitch, facing the opposition. They do not shake hands. They do not stretch. They wait for silence.

Then, the leader screams.

"Ka mate! Ka mate!"

The team screams back. "Ka ora! Ka ora!"

They stomp the ground in perfect unison. They slap their chests and thighs with enough force to leave red welts. They bulge their eyes and stick their tongues out to the root. This is the Haka. It is a traditional Māori challenge. It is technically a dance, but functionally, it is a threat display.

If you are standing on the other side of the halfway line, watching fifteen men who weigh 250 pounds each synchronizing their violence, your heart rate changes. Your cortisol spikes. Your lizard brain, the part of you that evolved to survive in the jungle, registers a massive, coordinated predator.

The Haka is not performed for the audience. It is performed for the enemy. It is a psychological weapon designed to win the game before the whistle blows. It hacks the mirror neurons of the opposition. When you see someone acting with extreme aggression and total synchronization, your brain simulates that aggression. You feel the threat physically. You feel small.

This is the first rule of weaponized sound: If you control the volume, you control the dominance hierarchy.

In the animal kingdom, size equals volume. A lion roars louder than a house cat. A bull elephant trumpets louder than a mouse.

Low-frequency, high-volume sound is a universal biological signal for mass. When a group of humans vocalizes in unison, they create a "super-organism" effect. They stop sounding like fifteen individuals and start sounding like one massive beast.

The acoustic footprint of the All Blacks doing the Haka is terrifying because it is disciplined. Noise is chaos; the Haka is order. It signals to the enemy: "We are not a mob. We are a single unit. We move together. We breathe together. If you fight one of us, you fight all of us."

This is the foundation of military discipline. Since the dawn of organized warfare, armies have used sound to intimidate enemies and regulate their own fear. The Romans slammed their swords against their shields. The Scots played bagpipes, an instrument designed specifically to sound like a dying animal screaming across the moors. The sound was meant to curdle the blood of the opposition.

But the primary use of military sound is not external; it is internal. It is about the "cadence."

Go to any Marine Corps boot camp in Parris Island or San Diego. You will hear the sound of the "Jody call."

"I don't know but I've been told!"

"Eskimo pussy is mighty cold!"

The lyrics are often profane, violent, or sexually explicit. They are designed to be offensive. But the content is irrelevant. The function is respiratory.

A platoon of recruits is a collection of civilians. They are out of shape, scared, and individualistic. The drill instructor has to turn them into a machine that can march twenty miles with a sixty-pound pack without stopping. He does this with the voice.

When the platoon sings a cadence, they are forced to exhale on the left foot strike. "Left, right, lo-right-a-lay-o." Every time the left boot hits the pavement, the air leaves the lungs.

This does three things mechanically.

First, it regulates the breathing. When humans are tired, we pant. We take shallow, chaotic breaths. This is inefficient. It leaves carbon dioxide in the blood, which leads to acidosis and fatigue. Singing forces a deep, rhythmic exhalation. It purges the CO2. It forces the recruits to breathe from the diaphragm. It creates a "gas exchange" efficiency that allows them to march further than they could in silence.

Second, it creates entrainment. We discussed this with the mother and infant, but here it is applied to locomotion. When you sing to a beat, your motor cortex locks onto the rhythm. You stop thinking about your legs. The movement becomes automatic. The "pain" signal from the feet is drowned out by the auditory signal from the throat. The recruits enter a trance state. They are not walking; they are being carried by the rhythm.

Third, it dissolves the ego. You cannot sing a cadence and think about your own problems at the same time. The vocalization requires focus. You have to listen to the caller and repeat the line instantly. You become part of the call-and-response loop. You stop being "Private Jones" and become "The Platoon."

This is why the military is so strict about volume. "Sound off!" the sergeant screams. "I can't hear you!" He is not deaf. He is checking for compliance and cohesion. A quiet platoon is a tired platoon. A loud platoon is a dangerous platoon. If you have the energy to scream, you have the energy to fight.

The psychological impact of this "shared voice" is profound. In the chaos of combat, the visual field is often obscured by smoke, dust, or terrain. You cannot see your unit. But if you can hear them—if you can hear the specific, rhythmic shout of your tribe—you know you are not alone. The sound is the tether.

This dynamic scales up from the platoon to the stadium.

Modern warfare is rare for the average civilian, but we simulate it every weekend in sports arenas. A football stadium is a tribal battleground. You have two armies (the teams) and two populations (the fans). The goal is territorial conquest.

The "Home Field Advantage" is a statistical reality. Teams win more often when they play at home. Analysts often attribute this to travel fatigue or sleeping in their own beds. But the primary factor is acoustic dominance.

Seattle's Lumen Field is designed to be a weapon. The roof is curved to reflect sound back onto the field. The seats are aluminum, which vibrates. When the opposing team is on offense, the fans—known as the "12th Man"—scream. They generate noise levels of 137 decibels.

For context, a jet engine at takeoff is 140 decibels. Pain begins at 125. Permanent hearing damage begins at 85.

The fans are not just cheering; they are jamming the enemy's communications. The opposing quarterback cannot hear the coach in his headset. He cannot scream audible changes to the offensive line. The players cannot hear the snap count. They have to rely on silent hand signals, which are slower and harder to see.

The noise physically disrupts the cognitive processing of the enemy. It induces stress. It causes false starts. It forces the offense to burn timeouts. The fans are literally changing the mechanics of the game using nothing but air pressure.

Meanwhile, the defense—the home team—feeds on it. The noise triggers an adrenal response. It is a "war cry" from 60,000 people backing them up. They rush faster. They hit harder. The sound creates a feedback loop of aggression.

European soccer takes this a step further with the "chant." Unlike the incoherent roaring of American football, soccer crowds sing organized songs. They have melodies. They have verses. They have conductors (the "capos" with megaphones).

When Liverpool FC plays at Anfield, the crowd sings "You'll Never Walk Alone." It is a slow, mournful ballad. It is not aggressive; it is religious. It signals absolute loyalty. It tells the players, "We are suffering with you. We are celebrating with you. We are family."

For the opposing team, hearing 50,000 people sing a hymn of loyalty to the other side is demoralizing. It reinforces the fact that they are intruders. They are in hostile territory. The sheer coordinated beauty of the song makes the hostility sharper. It says, "We are a culture. You are just visitors."

But sound does not have to be organized to be a weapon. It just has to be persistent.

This brings us to the dark side of the spectrum: Sonic Warfare and PsyOps (Psychological Operations).

The CIA and the US military have long understood that sound is an effective method of breaking human resistance without leaving physical marks. You can interrogate a prisoner with noise. You can break a siege with a playlist.

In 1989, the US Army surrounded the Vatican Embassy in Panama City. Inside was the dictator Manuel Noriega. He refused to surrender. The Army didn't want to storm the building and risk a diplomatic incident. So, they set up massive loudspeaker walls—the kind used for rock concerts—and aimed them at the windows.

They hit play.

They blasted "Panama" by Van Halen. "Welcome to the Jungle" by Guns N' Roses. "I Fought the Law" by The Clash. They played it at ear-splitting volumes, 24 hours a day.

Noriega was an opera fan. He hated rock music. But the taste wasn't the point. The point was the denial of silence.

The human brain requires silence to process information. It requires downtime to organize thoughts. When you are subjected to constant, high-decibel noise, you cannot think. You cannot sleep. Sleep deprivation induces psychosis within three days. The brain starts to hallucinate. Paranoia sets in. The will to resist crumbles because the cognitive energy required to maintain the resistance is being burned up trying to filter out the noise.

Noriega surrendered after 10 days. He walked out in his underwear. He was broken not by bullets, but by Van Halen.

We saw this again in Waco, Texas, during the siege of the Branch Davidian compound. The FBI blasted the sounds of rabbits being slaughtered, Tibetan chants, and Nancy Sinatra's "These Boots Are Made for Walkin'."

The randomness is part of the torture. If the noise is predictable, the brain can eventually tune it out. You can sleep through a train passing if it passes every night at 2 AM. But if the playlist switches from a crying baby to heavy metal to a Gilbert and Sullivan musical, the brain cannot habituate. It stays in a state of high alert. It keeps waiting for the next threat.

This technique was refined in the War on Terror. At Guantanamo Bay and CIA black sites, detainees were subjected to "music torture." They were shackled in stress positions in dark rooms and forced to listen to songs like "The Real Slim Shady" by Eminem or the "Meow Mix" commercial jingle on an infinite loop for days.

This sounds funny to a civilian. It sounds like a prank. It is not a prank.

Music, when stripped of choice, becomes a violation. Music is deeply tied to our emotional centers. It triggers memory and association. To have a piece of music forced into your ear canal against your will is a form of acoustic rape. It invades the interior space of the skull. You cannot close your ears like you can close your eyes. You are defenseless against the vibration.

The repetition weaponizes the "earworm" effect. The song loops in the brain even when the speakers are off. The prisoner cannot find a quiet place in their own mind. The jingle becomes the only reality. It erases the self.

Interrogators have reported that hardened terrorists, men who could withstand beatings and waterboarding, would break down and weep after 24 hours of the "Barney" theme song ("I love you, you love me..."). The juxtaposition of the childish, happy melody with the grim reality of the cell creates a cognitive dissonance that shatters the psyche. It makes the prisoner feel insane.

The biology is simple: The body reacts to loud noise as a physical threat. The sympathetic nervous system fires. Adrenaline pumps. If the noise never stops, the adrenaline never clears. The body enters a state of toxic stress. The organs begin to fail. The mind snaps.

This is the "Weaponized Frequency."

We are currently developing technology to do this without music. The Long Range Acoustic Device (LRAD) is a "sound cannon." It looks like a satellite dish. It can project a beam of sound up to 3,000 meters.

It can broadcast a voice message, but its primary setting is the "deterrent tone." This is a high-pitched, piercing squeal that hits the resonant frequency of the human ear canal. It causes instant, blinding pain. It causes nausea, dizziness, and disorientation. It triggers the flight reflex immediately.

Police use it for crowd control. Ships use it to repel pirates. It is a non-lethal weapon that uses the mechanics of the human ear against the owner. It reminds us that the ear is a vulnerability. It is a hole in the head that leads directly to the brain stem.

The effectiveness of weaponized sound proves the central thesis of this book: Sound is not air; it is matter. It is a physical force that interacts with biology.

When we sing together in a choir, we use that force to build. We use it to harmonize, to connect, to heal. We vibrate the air to create a structure of safety.

But the mechanics are neutral. The same vibration that heals can destroy. The same synchronization that builds a community can build an army. The same volume that creates a "safe space" in a church creates a "kill zone" in a stadium.

This duality is why the voice is so heavily regulated in authoritarian regimes. Dictators understand the power of the chant. They understand that if the people start singing their own songs, they are building a rival power structure.

This is why the Nazis placed such a heavy emphasis on singing. The Hitler Youth did not just march; they sang. They sang specifically crafted songs about blood, soil, and loyalty. The act of singing together physically wired the youth into the Nazi ideology. It bypassed their logic centers and went straight to the emotional core. It made them feel like a single organism—one that could kill without conscience because the "individual" had been dissolved into the "voice of the Volk."

You cannot fight a singing army with logic. You can only fight it with a louder song.

The "Baltic Singing Revolution" of the late 1980s is the counter-example. Estonia, Latvia, and Lithuania were under Soviet occupation. They had no weapons. They had no tanks. But they had a tradition of massive choral festivals.

In 1988, hundreds of thousands of Estonians gathered in Tallinn. They didn't riot. They sang. They sang the forbidden national anthems. They sang traditional folk songs. The Soviet soldiers stood there with their guns, confused. You can shoot a rioter. Can you shoot a choir? Can you shoot 300,000 people standing still, holding hands, and weeping while they sing a hymn about a swallow bird?

The optics were impossible. The sonic wall was impenetrable. The singing de-legitimized the tanks. It showed that the Soviet power was hard and brittle, while the Estonian power was soft and fluid.

The Soviets eventually left. The revolution was won without a drop of blood. It was won by the sheer, overwhelming weight of the human voice.

This is the ultimate lesson of weaponized sound. It is an arms race.

We are constantly bombarded by sonic weapons. The siren of the police car. The jackhammer. The aggressive bass of the car next to you at the stoplight. The relentless, scientifically optimized pop music in the grocery store designed to make you shop slower.

We live in a battlefield of frequencies.

If you are silent, you are a casualty. You are absorbing the noise of the world without pushing back. You are letting the environment dictate your internal rhythm.

To sing is to fight back. To sing is to claim the airspace. It is to say, "My frequency is stronger than your frequency."

When you walk down a dark street and you hum a tune, you are deploying a counter-measure. You are occupying your own auditory cortex so the fear cannot get in. You are creating a shield.

When you sing in the shower, you are claiming the territory of the bathroom. When you sing in the car, you are claiming the territory of the commute.

You are the All Black on the pitch. You are the Marine on the march. You are the fan in the stands.

You are not just making music. You are establishing dominance over the silence. You are warning the predators—both the lions in the grass and the demons in your head—that you are big, you are loud, and you are not to be eaten.

Chapter 8

The Hive Mind (Social Cohesion)

In 1665, a Dutch physicist named Christiaan Huygens was lying in bed, recovering from a minor illness. He was bored. To pass the time, he watched two pendulum clocks hanging on the wall across the room. He had invented the pendulum clock a few years earlier, so he knew the mechanics intimately. He noticed something odd.

The pendulums were swinging in perfect opposition. When one swung left, the other swung right. They were locked in an anti-phase synchronization. Huygens was suspicious. He assumed he had just started them that way by accident. So, he got out of bed and disrupted the rhythm. He stopped one clock, restarted it at a random interval, and waited.

Within thirty minutes, the clocks had returned to perfect synchronization.

He tried it again. He moved the clocks to different walls. The synchronization stopped. He moved them back to the same wooden beam. The synchronization returned.

Huygens realized that the clocks were communicating. As the heavy pendulum swung back and forth, it sent a tiny, imperceptible vibration through the wood of the clock case, into the beam of the wall, and into the other clock. The two machines were exchanging energy. It was physically easier for them to move together than to move apart. The path of least resistance was unity.

This phenomenon is called entrainment. It is a fundamental law of physics. If you put two oscillating systems in proximity and allow them to exchange energy, they will eventually synchronize.

It happens in the mechanical world, and it happens in the biological world. In the mangrove swamps of Southeast Asia, species of fireflies gather in the trees by the millions. At dusk, they blink randomly. But as the night deepens, the chaos resolves. One cluster synchronizes. Then the next tree joins in. By midnight, an entire mile of riverbank is flashing on and off like a single LED bulb.

They are not deciding to do this. They are not following a leader. They are visually entraining. The flash of a neighbor triggers the neural impulse of the observer. The gap between the flashes narrows until it hits zero.

Human beings are biological pendulums. We oscillate. We have a heart rate. We have a respiratory rate. We have brain waves. We have a walking cadence. And, like the clocks on the wall, when we get close to each other, we try to sync up.

If you walk down the street next to a friend, your strides will match within a minute. If you sit in rocking chairs on a porch, you will eventually rock at the same tempo. This is unconscious social glue. But the most aggressive, efficient, and total form of human entrainment is singing.

Singing overrides the subtle drift of individual biology and forces a hard reset of the entire group. It creates a "Hive Mind" not in a metaphorical sense, but in a physiological one.

The mechanism for this is the breath.

Most of our autonomic functions are automatic. You cannot choose to stop your kidneys from filtering blood. You cannot choose to stop your heart. But breathing is a hybrid. It is autonomic (you breathe when you sleep), but it is also somatic (you can hold your breath). It is the only handle we have on the nervous system.

There is a direct mechanical link between the lungs and the heart called Respiratory Sinus Arrhythmia (RSA). When you inhale, the diaphragm moves down, creating space in the chest. This lowers the pressure in the thoracic cavity. To compensate for the pressure drop, the heart beats faster. When you exhale, the diaphragm moves up, pressure increases, and the heart slows down.

Inhale, speed up. Exhale, slow down. This wave happens every minute of your life.

In a room full of people sitting in silence, this wave is chaotic. Person A is inhaling; Person B is exhaling. Their hearts are beating at different rates. Their nervous systems are isolated.

Then, they start to sing.

A song dictates a specific breathing pattern. You must inhale at the start of the phrase. You must exhale slowly and steadily to sustain the notes. You must pause at the comma. The melody acts as a rigid template for respiration.

In 2013, a study at the University of Gothenburg in Sweden measured this effect. They wired up a choir of high school students with heart rate monitors. They asked them to hum, to sing a hymn, and to chant a mantra.

The data was absolute. As soon as the singing started, the heart rate variability of the singers aligned. When the choir inhaled together, everyone's heart rate spiked. When they exhaled for the long phrase, everyone's heart rate dropped.

The graph showed a perfect sine wave. The hearts of fifteen individuals were beating as if they belonged to one body. The "clocks" had synchronized.

This is the "Choir Effect." It creates a state of physiological coherence. For the duration of the song, the boundary between "me" and "we" is physically erased. My heart is doing exactly what your heart is doing. My lungs are filling at the exact moment yours are filling. We are running the same software on the same hardware at the same time.

This synchronization explains the feeling of "ego dissolution" that singers report. The ego—the sense of being a distinct, separate individual—is an expensive energy drain. The brain spends a massive amount of glucose maintaining the construct of the "Self." It constantly monitors social status, safety, separation, and difference. "Am I better than him? Is she looking at me? Do I look stupid?"

This monitoring happens in the prefrontal cortex. It is the voice of judgment.

When you enter a state of extreme entrainment, the brain can stop generating the Self. The energy required to maintain the separation is higher than the energy required to join the flow. So, the brain lets go. The prefrontal cortex quiets down.

This is experienced as a profound relief. For a few minutes, you don't have to be you. You don't have to carry the weight of your own history, your own neuroses, or your own deadline. You are just a vibrating component of a larger machine.

This is why people join choirs. It is rarely about the music. It is about the drug of belonging. In a world that hyper-fetishizes individualism, where we are told to "build our brand" and "stand out," the act of disappearing into a harmony is a radical corrective. It is the only place where blending in is the goal.

The survival utility of this is obvious. A tribe that acts as a single organism is harder to kill.

Consider the murmur of starlings. Thousands of birds fly in a shifting, liquid cloud. They turn instantly. They dive instantly. They do not crash into each other. A hawk trying to attack the flock is confused. It cannot target a single bird because the target is constantly shifting. The "hive" protects the individual.

Humans are not birds, but we are herd animals. We survive in groups. The problem is that group living requires trust. You cannot sleep soundly if you think the person next to you might steal your food or slit your throat. You need a way to verify loyalty.

Singing is the verification algorithm.

In the evolutionary past, producing a complex, synchronized rhythm was a high-fidelity signal of cooperation. If I can sing with you, I am listening to you. I am matching your pitch. I am adjusting my volume to fit yours. I am demonstrating—in real-time—that I am willing to subvert my own desire for dominance to serve the group pattern.

You cannot fight and sing at the same time. The motor commands are incompatible. Singing triggers the parasympathetic nervous system (safety), while fighting triggers the sympathetic nervous system (danger).

Therefore, if a group of strangers meets and sings together, they are hacking the trust mechanic. They are signaling safety. This is known as the "Icebreaker Effect."

Corporate retreats use this, often badly. They force employees to drum or chant. It feels awkward because it is forced. But in organic settings—a football match, a church service, a protest march—the bonding is instantaneous.

Studies show that people who sing together cooperate better in subsequent tasks. If you take two groups of people, have Group A sing together and Group B listen to music together, and then give them a "Prisoner's Dilemma" game where they have to choose to share money or steal it, Group A shares significantly more.

The singing primed the cooperation pathways. It established a baseline of "we are on the same team."

This "Hive Mind" capability is likely why religion and music are inextricably linked. There is no religion on earth that does not use vocalization. The chant, the hymn, the call to prayer. The theology varies, but the method is constant.

The priests and shamans understood that you cannot build a congregation with logic. You can explain the dogma for hours, and people will nod politely. But if you get them to sing for ten minutes, they will feel the unity. They will feel the "spirit" moving through the room.

That spirit is just entrainment. It is the visceral sensation of one hundred heart rates locking into phase. It feels supernatural because it is one of the only times we physically feel the presence of others inside our own bodies. The vibration of the person next to you travels through the air, hits your eardrum, and resonates in your chest. You are literally being touched by their voice.

This physical intrusion is welcomed because it is structured. It is not noise; it is harmony.

Harmony itself is a lesson in social cohesion. In a choir, you are assigned a part. You are a Soprano, an Alto, a Tenor, or a Bass. The Sopranos get the melody. They are the stars. The Altos get the harmony—the weird, crunchy middle notes that don't sound like much on their own but give the chord its texture. The Basses provide the foundation.

If the Basses decide they want to be stars and sing the melody, the chord collapses. If the Sopranos decide to sing louder than everyone else, the balance is ruined.

Singing in harmony teaches you that your role is relative. You are essential, but you are not the whole. Sometimes you lead, sometimes you support. You have to listen louder than you sing.

This is a difficult lesson for the modern ego. We are trained to shout. We are trained to post. We are trained to broadcast our opinion regardless of the context. Social media is a choir where everyone is trying to sing a solo at the same time. The result is dissonance. It is noise.

A choir is the antidote to the comment section. In a choir, if you don't listen, you fail. If you are out of tune, you don't "disrupt the narrative"; you just ruin the song for everyone, including yourself. The feedback loop is immediate and physical.

This social binding effect is so strong that it works even when the participants hate each other.

There have been experiments in conflict zones—Northern Ireland, Israel and Palestine, Bosnia. Facilitators bring people from opposing sides into a room. These are people who have been taught from birth that the "other" is a monster. They cannot talk about politics. The conversation devolves into shouting within minutes.

So they stop talking. They build a choir.

It takes time. The body language is stiff. The eyes are averted. But the physics of entrainment does not care about ideology. It only cares about acoustics. As they learn the parts, as they are forced to breathe in sync, the biology takes over. The oxytocin begins to flow. The cortisol drops. The heart rates align.

By the end of the concert, the "monster" has become a "tenor." It is harder to kill a tenor. You have breathed with him. You have built a sound with him. The abstract hatred is replaced by a concrete, physical cooperation.

This doesn't solve the geopolitical crisis. It doesn't move borders. But it proves that the "Hive Mind" is a default setting that sits beneath the software of hate. The hate is a cognitive overlay. The harmony is a biological imperative.

We see the absence of this cohesion in the modern mental health crisis. We are the most individualized society in history. We live in single-person apartments. We drive single-person cars. We listen to music on noise-canceling headphones, ensuring that we share the acoustic space with absolutely no one.

We have broken the hive.

The result is a feeling of chronic exposure. A bee separated from the hive is an anxious bee. It buzzes erratically. It dies quickly. Humans separated from the group exhibit the same symptoms. Anxiety, depression, and a sense of existential dread are the predictable results of a social animal living in isolation.

We try to fix this with chemicals. We try to fix it with therapy. But we rarely try to fix it with physics. We rarely say, "Go stand in a room with fifty other people and vibrate your sternums together until your hearts sync up."

And yet, when people do it, they describe it as life-saving.

The "Pub Choir" movement in Australia and the "Choir! Choir! Choir!" movement in Canada are testing this. These are not formal choirs. There are no auditions. You show up at a bar. You get a beer. You get a lyric sheet. You learn a three-part harmony to a Bon Jovi song.

The participants are not singers. They are accountants, mechanics, nurses, and students. They look terrified at the start. But when the conductor gets the room locked in—when three hundred strangers suddenly hit a major chord—the energy shift is nuclear. You can see it in the footage. Faces open up. People cry. They high-five strangers.

They are experiencing the relief of the hive. They have plugged back into the mainframe.

The "Hive Mind" is often used as a derogatory term in science fiction. It implies a loss of free will. It implies drones serving a queen. But for a species that evolved on the savannah, surrounded by predators, the hive was the only safety we had.

We are not eagles. We are not tigers. We are soft, slow, and delicious. Our only defense was the group. The voice was the mechanism that solidified the group. It was the signal that kept the edges of the tribe defined.

When we sing "We Shall Overcome," or "Amazing Grace," or "Sweet Caroline," we are activating that ancient defense mechanism. We are asserting that we are many, and therefore we are safe.

The silence of the modern world is a signal of fragmentation. A silent crowd is a crowd of individuals, vulnerable and alone. A singing crowd is a fortress.

The physics are clear. Energy moves more efficiently when systems are synchronized. Resistance drops. Power increases.

If you want to go fast, go alone. If you want to go far, go together. But if you want to survive the night, you had better sing the same song.

Chapter 9

Vibrating the Vagus (Internal Physiology)

You have a cable running through your body.

It is the width of a shoelace. It originates in the medulla oblongata, the reptilian base of the brain, and snakes its way down through the neck, passes the vocal cords, wraps around the heart, weaves through the lungs, punches through the diaphragm, and terminates in the viscera of the gut. It is the longest of the cranial nerves. Anatomists call it Cranial Nerve X. The Latins called it the Vagus, meaning "The Wanderer," because it roams through the interior of the body like a root system.

This nerve is not just a passive wire. It is the command line for your internal state. It controls the parasympathetic nervous system, the "rest and digest" mode that counteracts the adrenaline-fueled "fight or flight" mode. When the Vagus is active, your heart rate slows. Your blood pressure drops. Your digestion turns on. Your immune system engages. You feel safe.

When the Vagus is blocked or inactive, you feel the opposite. The heart races. The stomach knots. The mind scans for threats. You are in a state of sympathetic dominance.

Most people spend their lives trying to manipulate this nerve from the outside. They take beta-blockers to slow the heart. They take antacids to settle the stomach. They drink wine to calm the nerves. They are trying to hack the software by smashing the hardware with chemistry.

But the Vagus has a physical access point. It has a button you can press. That button is located in the throat, and the finger you use to press it is the human voice.

To understand the mechanism, you have to look at the geography of the neck. The Vagus nerve exits the skull via the jugular foramen and runs down the carotid sheath, right alongside the carotid artery and the jugular vein. As it passes through the neck, it sends off branches. One of the most important branches is the Superior Laryngeal Nerve. Another is the Recurrent Laryngeal Nerve.

These nerves innervate the larynx. They control the muscles that move the vocal folds. They control the tension, the pitch, and the closure of the glottis.

Because the nerve fibers are physically woven into the muscle tissue of the voice box, the act of phonation—making sound—directly stimulates the nerve. When you hum, you are not just making a noise; you are creating a mechanical vibration that massages the Vagus nerve.

[Image of the vagus nerve anatomy]

This is not a metaphor. It is simple physics. Sound is a pressure wave. When you make a low, resonant sound in your throat, you create a standing wave that vibrates the tissues of the pharynx, the laryngeal cartilage, and the surrounding musculature. This vibration is transferred directly to the nerve sheath.

It stimulates the fibers. The signal travels up to the brainstem, telling the brain, "We are vibrating. We are exhaling. We are safe." The brain then sends a signal back down the main trunk of the Vagus to the heart, releasing acetylcholine, a neurotransmitter that acts as a tranquilizer. The heart rate drops.

This is the "Vagal Tone." It is a measure of how quickly your body can switch from stress to calm. High vagal tone means you bounce back instantly. Low vagal tone means you stay stressed for hours after the threat is gone.

Singing increases vagal tone. Specifically, singing low, long notes increases vagal tone.

This specificity matters. Not all sounds are created equal. If you scream in a high pitch, you are creating tension. You are tightening the laryngeal muscles, raising the larynx, and signaling distress. High-frequency sounds are associated with fear—the shriek of a monkey, the cry of a baby. These sounds trigger the sympathetic nervous system. They tell the brain to panic.

But low-frequency sounds do the opposite. Think about the sounds of safety in nature. The purr of a cat. The rumble of a mother bear. The distant roll of thunder. Low frequencies travel through the ground. They imply size. They imply stability.

When you produce a low, resonant tone—like a hum or a deep chant—you are mimicking the acoustic signature of a large, calm animal. You are tricking your own brain. You are signaling, "I am big. I am heavy. I am not being hunted."

This explains the universality of the "Om."

In the Dharmic traditions of India, the "Om" (or Aum) is considered the primordial sound of the universe. Spiritually, it is significant. But physiologically, it is a masterclass in vagal stimulation.

Analyze the phonetics. It starts with "Aaa" (open throat, vibration in the chest). It moves to "Ooo" (vibration moves up the throat). It ends with "Mmm" (vibration in the lips and nasal cavity, buzzing the entire skull).

The "Mmm" component is crucial. This is "Bhramari Pranayama," or "Bee Breath." When you hum with your lips closed, you trap the air pressure inside the oral and nasal cavities. This increases the intra-oral pressure. It turns the skull into a resonance chamber. The vibration is amplified.

If you put your hand on your chest while humming a low "Mmm," you can feel the rib cage rattling. That rattle is shaking the Vagus nerve where it wraps around the heart and lungs. You are literally vibrating your own anxiety to death.

Studies show that five minutes of "Bee Breath" significantly lowers blood pressure and heart rate. It is more effective than silent deep breathing. The sound adds the mechanical component that the breath alone lacks.

This is why we groan when we are in pain.

Watch a person with a stomach ache. Watch a person with the flu. Watch a person who has just received terrible news. They rock back and forth, and they make a sound. "Ohhhhh. Ughhhhh."

We often interpret this as a plea for help. We think they are communicating their suffering to others. But people do this even when they are alone. They do it when they are semi-conscious.

They are self-soothing. The body knows, on an instinctual level, that making a low, grinding noise helps manage the pain. The groan vibrates the chest. It stimulates the Vagus. It triggers the release of endorphins. It is an autonomic attempt to down-regulate the nervous system.

We train children out of this. We tell them, "Don't whine. Be quiet. Stop making that noise." We strip them of their natural analgesic tool. We force them to suffer in silence, which removes the vagal stimulation and leaves them trapped in the high-cortisol state of silent distress.

This mechanism is the core of Dr. Stephen Porges' "Polyvagal Theory." Porges identified that the Vagus system has two branches. The "Dorsal Vagal" is the primitive branch—it causes freezing and shutting down (playing dead). The "Ventral Vagal" is the evolved branch—it controls social engagement and safety.

The Ventral Vagal complex is linked to the muscles of the face and the voice. This is the "Social Nervous System." When we speak with "prosody"—melody and rhythm—we are exercising the Ventral Vagal system. We are keeping the pathway to safety open.

When trauma occurs, people often lose their voice. They go "flat." Their voice becomes monotone. They lose the ability to inflect. This is a sign that the Ventral Vagal system has gone offline. They are stuck in a survival state.

To heal, they have to turn the system back on. They have to re-introduce melody to the voice.

This is why chanting is a staple of every major religion. Gregorian chants, Tibetan throat singing, the Islamic Adhan, Hebrew cantillation. These are not just aesthetic choices. They are physiological technologies.

The monks didn't have MRI machines, but they had trial and error. They figured out that if you sit in a stone room and sing low notes for an hour, your mind stops racing. Your body settles. You enter a state of receptive contemplation.

They were engineering their own neurobiology. They were using the voice to pull the handbrake on the metabolic engine.

The "long exhalation" is the other half of the equation.

You cannot sing without controlling the exhale. To hold a note for ten seconds, you have to release the air slowly. As we established, exhalation is linked to the slowing of the heart.

When you are stressed, you pant. You take short inhales and short exhales. Or you hyperventilate—too much inhale, not enough exhale. This floods the blood with oxygen but depletes carbon dioxide, causing dizziness and panic.

Singing forces the ratio to flip. In a typical song, you might inhale for one second and exhale for five or ten seconds. This 1:10 ratio is a powerful signal to the brainstem. It says, "We cannot possibly be running from a tiger, because we are exhaling for ten seconds."

The brain accepts this data and lowers the threat level.

This creates a feedback loop. You feel a little calmer, so your voice gets a little steadier. Your voice gets steadier, so you vibrate the nerve more effectively. The nerve fires, and you feel calmer still.

This is the "upward spiral" of singing. It is the opposite of the panic attack's "downward spiral," where a racing heart scares you, which makes the heart race faster.

The physical sensation of resonance also provides a "somatosensory anchor." Anxiety is often described as being "in your head." It is a whirlwind of future-based thoughts. "What if I get fired? What if I get sick?"

Singing pulls you out of the future and into the body. You cannot ignore the buzzing in your lips. You cannot ignore the engagement of your abs. The sensation is immediate and local. It grounds you.

This is "interoception"—the awareness of the internal state of the body. People with high anxiety often have poor interoception. They try to ignore their body because it feels like a dangerous place. Singing forces them to inhabit the body safely. It turns the body from a cage into an instrument.

The volume of the sound matters as much as the pitch.

To vibrate the Vagus effectively, you need a certain amount of acoustic energy. Whispering doesn't do it. Whispering actually creates tension in the vocal folds without the resonance. It dries out the cords.

You need to "phonate" fully. You need the folds to close completely and the air to drive through them. This requires what vocal coaches call "support"—the engagement of the diaphragm.

When you engage the diaphragm, you are also massaging the Vagus nerve from the bottom up. The nerve passes through the diaphragm. The rhythmic contraction of the muscle stimulates the nerve fibers in the abdomen.

So, a full-voiced song attacks the nerve from two directions: the vibration in the throat and the movement in the belly. It is a pincer movement on stress.

This explains why "car singing" is so therapeutic. It is a private space. You are not worried about being heard. So you sing loud. You belt. You engage the whole system. You get out of the car feeling lighter, not because the traffic was good, but because you just gave yourself a twenty-minute vagal massage.

The tragedy is that we reserve this regulation for "happy" moments. We sing when we are happy. We should be singing when we are miserable.

We have a cultural script that says when you are stressed, you should be quiet. You should "calm down." But "calming down" via silence is hard. It requires a mental discipline that an anxious brain doesn't have.

It is much easier to "vibrate down."

This is relevant for people with chronic pain. The Vagus nerve plays a role in inflammation. High vagal tone suppresses the production of cytokines—proteins that cause inflammation. By stimulating the nerve, you can actually reduce the systemic inflammation in the body.

There are now devices on the market—Vagus Nerve Stimulators (VNS)—that jolt the nerve with electricity. They are implanted in the chest or clipped to the ear. They are used to treat drug-resistant epilepsy and severe depression. They work.

But you have a VNS installed in your throat. It is free. It is powered by air.

The "Om" works. The "Hum" works. The "Hymn" works.

The specific vowel sounds affect different parts of the body due to "sympathetic resonance." This is the idea that different frequencies vibrate different masses.

Low sounds (like "Uhh" or "Ooo") vibrate the chest.

Mid-range sounds (like "Ahhh") vibrate the throat and upper chest.

High sounds (like "Eeee") vibrate the head and sinuses.

If you feel a knot in your chest, low humming targets that mass. If you have a tension headache, a higher, nasal hum might target the blockage in the sinuses. You can learn to direct the sound like a laser.

This is not mysticism; it is acoustics. You are a container of water and bone. Sound waves travel through water four times faster than through air. You are highly conductive.

When you stop singing, you lose this internal maintenance tool. You become a stagnant pond. The vibration stops. The nerve goes dormant. The stress accumulates in the tissues with no release valve.

The modern "silent treatment" we give ourselves is a health hazard. We sit in silent offices. We sit in silent cars. We watch silent screens. The only noise is the noise of the world coming in. We rarely push noise out.

This creates an imbalance. We are absorbing, but not emitting.

To fix this, we need to reclaim the groan. We need to reclaim the hum. We need to normalize the idea that making a sound is a valid response to stress.

If you are sitting at your desk and you feel the panic rising, close your mouth and hum. A low, steady, buzzing hum. Do it for two minutes. Feel the chest vibrate. Feel the breath lengthen.

Watch what happens to your shoulders. Watch what happens to your jaw.

The Vagus nerve will pick up the signal. It will send the message to the heart. The heart will slow down. The tiger will disappear.

You didn't think your way out of the panic. You vibrated your way out. You used the machinery of the throat to hack the machinery of the brain.

This is the internal architecture of the howl. It is not just a broadcast to the world; it is a signal to the self. It is the feedback loop that keeps the machine from overheating.

The nerve is waiting. The vocal folds are ready. All you have to do is push the air and let it buzz.

Chapter 10

The Analgesic Scream (Pain Management)

Watch a professional tennis match with the volume up.

It sounds like a torture chamber. Every time the racket connects with the ball, the player emits a sound that is somewhere between a bark and a shriek. It is loud—often exceeding 100 decibels. It is guttural. It is ugly. For years, commentators and spectators complained that the "grunting" was annoying, unsportsmanlike, or a form of cheating designed to distract the opponent. They tried to ban it.

They failed, because the players weren't doing it for attention. They were doing it for physics.

A study conducted at Iowa State University analyzed the effect of grunting on grip strength and ball velocity. The results were not subtle. When players grunted, their hitting power increased by an average of 3.8 percent. That sounds small, but in a game where the ball travels at 120 miles per hour, 3.8 percent is the difference between an ace and a returnable volley.

The scream is a performance-enhancing drug.

The mechanism is the "Valsalva maneuver." When you vocalize sharply and loudly, your vocal folds slam shut for a microsecond before releasing the air. This creates a spike in intra-abdominal pressure. It turns your torso into a rigid cylinder. This rigidity provides a stable platform for the pectoral and shoulder muscles to fire against. You cannot fire a cannon from a canoe; you need stable ground. The scream solidifies the ground.

But the utility of the scream goes beyond power. It is primarily about pain.

Pain is not just a sensation; it is a signal. It travels from the nerve endings (nociceptors) up the spinal cord to the brain. The brain receives the signal and decides how much "suffering" to attach to it. This decision is not fixed. It can be modulated.

One of the most effective ways to modulate it is to make noise.

This is the "Analgesic Scream." It is the reason you instinctively shout "Ow!" or "Fuck!" when you stub your toe. You don't decide to do it. The vocalization bypasses the executive centers of the brain and fires directly from the motor cortex. It is a reflex.

If the vocalization were useless, evolution would have pruned it. A shout reveals your location to predators. If you are injured and you scream, you are telling the lion exactly where the easy meal is. The fact that we still scream when hurt suggests that the benefit of the noise outweighs the risk of predation. The benefit is that the noise actually turns down the volume of the pain.

Researchers at the National University of Singapore tested this. They asked participants to plunge their hands into a bucket of ice water—a classic torture test known as the "Cold Pressor Task." They measured how long the subjects could keep their hands submerged.

In the first round, the subjects had to stay silent. In the second round, they were allowed to say "ow" or scream when it hurt.

When they vocalized, they could withstand the pain for nearly 30 percent longer.

The theory is "Gate Control." The spinal cord is like a gate. It can only process a certain amount of information at once. The pain signal is traveling up the thin, slow nerve fibers (C-fibers). The vocalization triggers a massive burst of sensory input—the vibration of the chest, the movement of the jaw, the sound in the ears. This new data travels on faster, thicker nerve fibers. When this flood of data hits the spine, it jams the gate. The pain signal gets crowded out.

Essentially, you are distracting your own nervous system with a louder signal. You are D DoS-ing the pain server.

This explains why the specific word you choose matters.

In a separate study at Keele University in the UK, researchers compared the analgesic effects of swearing versus shouting a neutral word. One group put their hands in ice water and shouted "Fuck!" repeatedly. The other group shouted "Table!"

The swearers won. They held their hands in the ice significantly longer and reported a lower perceived pain intensity.

The neutral word "table" engages the language centers, but it doesn't engage the emotional centers. A curse word is different. It is stored in the amygdala and the basal ganglia—the primitive, emotional brain. When you shout a taboo word, you trigger a "fight or flight" response. This dumps adrenaline into the blood. Adrenaline is a powerful painkiller. It is the body's combat drug.

So, shouting "Fuck!" is not just an expression of anger; it is a self-administered shot of epinephrine.

This brings us to the ultimate arena of pain and vocalization: Childbirth.

Labor is the most intense physical pain a human can experience without dying. It involves the dilation of the cervix from zero to ten centimeters and the violent contraction of the uterus, a muscle the size of a pear trying to push a watermelon through a specific bone channel.

In modern hospitals, the default approach to this pain is the epidural. We numb the nerves from the waist down. We silence the body.

But for 200,000 years, women didn't have epidurals. They had the "Labor Song."

Midwives in traditional cultures do not tell women to be quiet. They encourage sound. But they encourage a specific kind of sound.

When a woman is in early labor, the pain is manageable. She might chat or breathe quietly. As the contractions intensify, the instinct is to panic. Panic triggers the sympathetic nervous system. The breath becomes shallow and high in the chest. The throat tightens. The woman starts to produce high-pitched, thin screams.

"High and tight" is the enemy of birth.

There is a physiological connection between the jaw and the pelvis. This is known as the "Sphincter Law." When the jaw is tight, the pelvic floor is tight. When the throat is constricted, the cervix is constricted. If a woman is screaming in a high-pitched panic, she is physically fighting the opening of her own body. She is clamping down on the baby.

The midwife's job is to drop the pitch.

She will look the woman in the eye and model a low, guttural moan. "Ohhhhhhh." "Ahhhhhhh."

She encourages the woman to open her jaw, relax her lips (often by blowing "horse lips"), and direct the sound down into the belly.

When the woman switches from a high scream to a low roar, the physiology changes. The jaw unlocks. The pelvic floor relaxes. The vibration of the low note travels down the spine and helps numb the sacrum. The deep, forced exhalation required to make the sound ensures that the baby gets oxygen.

This low vocalization is a form of trance induction. The rhythm of the moan often matches the rhythm of the contraction. The woman rides the wave of the pain using the sound as a surfboard. If she stops making the sound, she falls off the board and drowns in the sensation.

We see this "vocal anesthesia" in the "Transition" phase—the most intense part of labor right before pushing. Women often enter an altered state of consciousness. They stop speaking in sentences. They become primal. They growl. They roar.

This is not "drama." It is the brain shutting down the neocortex (logic) and letting the brainstem drive the car. The noise is the engine of that drive.

The modern medical model often misunderstands this noise. Nurses might tell a laboring woman to "hush" or "save her energy." This is bad advice. Suppressing the vocalization suppresses the release of endorphins. It forces the woman to bottle the energy of the contraction inside her body, which creates tension, which increases pain.

Silence amplifies suffering.

This is true in the hospital, and it is true in the gym. It is true in the dentist's chair.

Consider the breath. Pain causes a "breath holding" reflex. When you anticipate a needle stick or a heavy lift, you instinctively stop breathing. This is a bracing mechanism. It protects the core, but it traps carbon dioxide and spikes blood pressure. It tells the brain, "Threat imminent."

Singing—or sustained vocalizing—forces the breath to move. You cannot hold your breath and scream at the same time.

The act of pushing air out is a signal of release. It tells the parasympathetic system that the threat is being managed. It keeps the blood oxygenated.

This is why martial artists use the "Kiai." When a karate player strikes, they shout "Hi-yah!" (or a specific syllable). They function exactly like the tennis player. They are tightening the core for impact, but they are also exhaling to prevent the wind from being knocked out of them if they get hit.

The sound serves as a shield.

The relationship between sound and pain is also about agency. Pain is passive. It happens to you. It is an invasion.

Vocalization is active. It is something you do. It is an output.

When you scream, you are reclaiming agency. You are answering the invasion with an expulsion. You are turning the passive experience of suffering into an active experience of expression.

This psychological shift is crucial. Feeling helpless increases the perception of pain. Feeling like you are "doing something" decreases it. The scream is the "doing."

This is why we see "primal scream therapy" endure in various forms. In the 1970s, Arthur Janov popularized the idea that screaming could release repressed childhood trauma. While the specific Freudian mechanics of his theory have been largely debunked, the physical utility remains.

People who go into the woods, or into a soundproof car, and scream at the top of their lungs report a feeling of lightness afterward. They feel "cleared out."

They have completed the stress cycle.

Stress is a biological sequence. It has a beginning, a middle, and an end. In the wild, the lion chases you (stress begins). You run (stress peaks). You escape (stress ends). The physical exertion of running burns off the cortisol and signals the end of the cycle.

In modern life, the stress (the email, the divorce, the traffic) hits us, but we don't run. We sit still. We don't scream. We stay polite. The cycle never closes. The chemicals stay trapped in the muscles.

The scream is a synthetic way to close the loop. It mimics the physiological peak of the "fight." It dumps the energy. It allows the body to reset.

We see the absence of this tool in the way we handle chronic pain. Patients are often told to "manage" their pain quietly. We view vocalization as a sign of weakness or a loss of dignity.

"He suffered in silence." We write this on tombstones as a compliment. We consider stoicism to be the highest virtue.

But biologically, stoicism is expensive. It requires massive inhibition. You have to use your prefrontal cortex to override the impulses of the limbic system. You are fighting a war on two fronts: the pain itself, and the effort to hide the pain.

This doubles the metabolic load. It exhausts the patient.

If we allowed people to make noise—to hum, to groan, to chant—they might actually suffer less.

We can look at the "Keening" traditions of Ireland or the professional mourners of the Middle East as a social application of this.

Grief is a form of physical pain. Heartbreak activates the same neural regions as a broken leg.

Keening is a high-pitched, wailing song performed over the dead. It is piercing. It is designed to be disturbing. It forces the grief out of the body. It prevents the mourner from freezing up.

The wail is a pressure valve for the community. It says, "The pain is here. It is loud. We are not hiding it."

By externalizing the pain, they prevent it from becoming internal trauma.

In our sanitized culture, we sedate the grief. We give the widow Valium. We keep the funeral quiet. We ask everyone to be "respectful," which means "silent."

We are denying the analgesic scream. We are forcing the pain to stay inside, where it rots.

The "Analgesic Scream" is not just about volume. It is about vibration.

There is a theory called "Vibroacoustic Therapy." Low-frequency sound vibrations (between 30 Hz and 120 Hz) can penetrate the body and relax muscle tissue. They can increase blood flow and reduce spasms.

When you hum a low note, you are creating these frequencies internally. You are essentially sitting on a massage chair that is located inside your own throat.

If you have a tension headache, the muscles of the neck and scalp are contracted. Humming creates a mechanical vibration that shakes these muscles loose. It is physical therapy from the inside out.

Cats do this. The "purr" is not just a sign of happiness. Cats purr when they are injured. They purr when they are giving birth. The frequency of a cat's purr (between 25 and 150 Hz) is exactly the frequency that promotes bone healing and tissue regeneration.

The cat is using its voice to heal its own broken bones. It is a biological ultrasound machine.

Humans don't purr, but we can hum. We can groan. We have the machinery. We just refuse to turn it on because we are afraid of looking weird.

This fear of "looking weird" is the muzzle.

Imagine a man with a kidney stone. He is curled up on the floor. He wants to scream. But he is worried that the neighbors will call the police. He is worried that his wife will be frightened. So he bites a towel. He clamps his jaw.

He has just voluntarily discarded his most potent painkiller. He has chosen social decorum over biological relief.

This is the modern condition. We are a species that is in pain, but we are terrified of the sound of pain. We want the world to be frictionless and quiet.

But the body is not quiet. The body is a noisy, messy engine. Birth is noisy. Death is noisy. Sex is noisy. Effort is noisy.

When we sanitize the soundtrack of our lives, we sanitize the experience. We detach from the reality of the flesh.

To reclaim the analgesic scream is to accept that you are an animal. It is to accept that when the needle goes in, or the weight comes down, or the heart breaks, you are allowed to roar.

You are allowed to use the air to push back against the pressure.

The next time you are in pain, try it. Don't be polite. Don't bite your lip.

Open your mouth. Drop your jaw. Engage your diaphragm.

And make the sound that the pain wants to make.

It might be a grunt. It might be a wail. It might be a curse word that would make a sailor blush.

Let it out.

Feel the vibration in your chest. Feel the gate in your spine slam shut against the signal. Feel the adrenaline wash the system.

You will still hurt. But you will not be a victim of the hurt. You will be a participant.

And that difference—the difference between enduring and expressing—is the difference between breaking and bending.

Chapter 11

Metabolizing Misery (The Blues & Grief)

Adele releases a ballad about a breakup. It is four minutes of pure, distilled misery. She describes the heartbreak in excruciating detail. The piano is somber. Her voice cracks with genuine emotional weight.

The song goes to number one in thirty countries. It is played at weddings. It is played in cars during rush hour. Millions of people, many of whom are perfectly happy in their relationships, listen to it on repeat and weep.

This is the "Sad Song Paradox."

From a purely hedonistic perspective, this behavior is irrational. Organisms typically avoid pain. We do not put our hands on hot stoves. We do not eat rotten meat. Yet, we actively seek out art that induces sadness. We pay money to be made miserable.

The reason is that sadness is not the goal. The goal is the exit.

We listen to sad music—and more importantly, we sing sad music—because it is a mechanism for metabolizing grief. Grief is a heavy, dense biological state. It slows the metabolism. It depresses the immune system. It creates a feeling of physical weight in the chest. If it stays inside the body, it rots. It turns into depression.

To clear it, you have to move it. You have to liquefy the solid mass of sorrow and turn it into something that can flow out of the system. That flow is the song.

This process is what the Greeks called catharsis. The word is often misused to mean "emotional release," but its medical root is closer to "purgation." In ancient medicine, a cathartic was a drug that induced vomiting or cleared the bowels. It was a violent expulsion of toxins.

Singing the Blues is emotional vomiting. It is the act of taking the poison inside you—the heartbreak, the poverty, the injustice—and forcing it out through the throat.

The Blues, as a musical form, was not invented to make people sad. It was invented to keep people alive. It emerged from the Mississippi Delta, from a population living under the crushing weight of Jim Crow, poverty, and manual labor. These people did not have access to therapy. They did not have access to antidepressants. They had the twelve-bar progression.

Analyze the structure of a standard Blues verse. It is AAB.

Line A: You state the problem. ("I woke up this morning, blues all around my bed.")

Line A: You repeat the problem. ("I woke up this morning, blues all around my bed.")

Line B: You resolve or comment on the problem. ("I didn't have no breakfast, and the blues was all I had.")

The repetition is functional. The first line identifies the pain. "Here is the thing that hurts." The second line confirms it. "Yes, it really hurts." The repetition forces the singer to stay with the feeling, to look it in the eye.

But the third line is the pivot. The third line provides the rhyme and the resolution. It wraps the pain in a structure. By rhyming "bed" with "had," the singer exerts control over the chaos. They have taken a messy, terrifying life event and wrestled it into a box that has rhythm and meter.

This act of structuring is the metabolism.

When you are in the grip of raw grief, you feel helpless. The emotion is happening to you. It is a wave that is drowning you.

When you sing about the grief, you become the narrator. You step outside the wave and describe the wave. You move the processing of the event from the amygdala (the panic center) to Broca's area (the speech center) and the prefrontal cortex (the logic center). You turn "feeling" into "syntax."

This shift is analgesic. As soon as you can name the monster, the monster shrinks.

But the singing does something else: It triggers the release of Prolactin.

Prolactin is a peptide hormone. It is best known for stimulating milk production in lactating mothers. But it is also a major component of the stress response. When you are injured, or when you cry "emotional tears" (as opposed to the reflex tears from cutting an onion), the body releases Prolactin.

Prolactin is a tranquilizer. It soothes. It promotes a feeling of calm and consolation. It is the body's way of hugging itself after a trauma.

Studies show that singing sad songs—or listening to them with high empathy—triggers this Prolactin release. It tricks the body into thinking it has endured a trauma and survived, so it dispenses the comfort hormone.

This is why you feel better after a "good cry." You don't feel better because the situation has changed. You feel better because you are chemically sedated by your own tears.

Happy music doesn't do this. "Walking on Sunshine" might give you a dopamine hit of energy, but it doesn't console you. It doesn't touch the Prolactin system. To get the comfort, you have to go through the sadness.

This biological necessity explains the ancient tradition of Keening.

In pre-industrial Ireland, and in cultures across the Middle East and Africa, a funeral was not a quiet affair. It was a sonic event. When someone died, the family did not just sit in polite silence. They hired a "Keener" (in Irish, bean chaointe).

The Keener was usually a woman. She was a professional. Her job was to stand over the body and scream.

It wasn't a random scream. It was a highly stylized, melodic wail. It had a specific rhythm. It started low and rose to a fever pitch. It involved rocking the body, tearing the hair, and beating the chest.

The Keener's function was to act as a lightning rod for the community's grief. Death is shocking. It induces numbness. The family is often in a state of freeze. They cannot process the reality of the loss.

The Keener breaks the freeze. By performing the grief at maximum volume, she gives everyone else permission to collapse. She pierces the numbness with a sound that is so agonizing, so raw, that it forces the sympathetic resonance of the room.

If you hear someone keening, you cannot stay stoic. The sound vibrates your own chest. It triggers your own mirror neurons. You start to cry.

This is the "infection" model of healing. The Keener infects the room with the expression of pain so that the room can purge the pain. She is drawing the poison out of the wound by sucking on it.

Once the wailing is done, the energy changes. The heavy, suffocating atmosphere of the wake lifts. The tears have been shed. The Prolactin has been released. The community can now move to the next stage: telling stories, drinking whiskey, and celebrating the life.

You cannot get to the celebration without the wail. If you skip the wail, the grief stays stuck.

In modern Western culture, we have banned the wail. We consider it undignified. We consider it hysterical. We medicate the widow so she doesn't make a scene. We hold the funeral in a carpeted room with hushed tones. We play recorded organ music.

We have sanitized death, and in doing so, we have broken the mechanism of recovery.

We leave the funeral with the grief still trapped in our tissues. We carry it home. It turns into chronic anxiety. It turns into ulcers. It turns into insomnia. Because we didn't metabolize it, it is still sitting in the stomach like a stone.

The "Blues" is a secular form of Keening. It is a way to wail without a dead body in the room.

When B.B. King sings "The Thrill Is Gone," he is keening for a dead relationship. He is making the guitar cry because he is too grown to scream. But the note bending on the guitar string mimics the human wail perfectly. It hits the same frequency. It triggers the same release.

This brings us to the concept of "Validation."

Grief is isolating. When you are in deep pain, you feel like you are the only person on earth who has ever felt this way. You feel crazy. You feel broken. The world keeps turning—people are buying coffee, the sun is shining—and it feels like an insult.

When you hear a song that describes your specific pain, the isolation shatters.

"I am not crazy. This person felt it too. They felt it so strongly that they wrote it down, recorded it, and now they are singing it to me."

The song acts as a handshake across the void. It validates the reality of the suffering. It says, "Your pain is human. It is common. It is survivable."

Singing along intensifies this connection. When you sing the lyrics, you are physically inhabiting the experience of the songwriter. You are trying on their grief. If they survived it, you can survive it.

This is why anthems of misery—like "I Will Survive" or "Someone Like You"—are so powerful in group settings. Go to a karaoke bar. Watch what happens when a heartbreak anthem comes on. The drunk girl in the corner isn't singing it alone. The whole room sings it.

They are not singing to the screen. They are singing to each other. They are acknowledging the universal reality of loss. "We have all been dumped. We have all been fired. We have all lost someone."

The chorus becomes a support group. For three minutes, the shame of the failure is removed. The failure is transmuted into art.

This transmutation is the alchemy of singing. You take lead (misery) and turn it into gold (music).

Physiologically, this involves the breath. Grief affects the respiratory system. It causes the "sob."

A sob is a specific breathing pattern: a sharp, spasmodic inhalation followed by a stuttering exhalation. It is the diaphragm spasming. It is the body trying to get oxygen while hyperventilating from distress.

Singing overrides the sob. You cannot sob and sustain a note at the same time. The muscular control required to hold a pitch forces the diaphragm to smooth out. It resolves the spasm.

When you sing through tears, you are physically wrestling your respiratory system back under control. You are using the structure of the song to splint the broken bone of the breath.

This is why we sing "We Shall Overcome" at protests, even while being beaten or arrested. The singing prevents the panic sob. It keeps the air moving. It keeps the line holdable.

The "Blues" mindset also teaches us something crucial about the nature of happiness. The Blues does not promise a solution. It does not promise that the woman will come back, or that the boss will give you a raise, or that the rain will stop.

The Blues promises endurance.

"The sun's gonna shine in my back door someday."

It acknowledges the misery of the present while insisting on the possibility of a future. It is not optimism; it is stamina.

By singing the complaint, you are exercising your stamina. You are proving to yourself that the pain did not kill you. You had enough breath left to sing about it.

This is the "aesthetic distance." If you can sing about the trauma, you are no longer in the trauma. You are observing it. You have gained a few inches of separation.

In the brain, this looks like a shift in blood flow. When you are drowning in emotion, the limbic system is lit up like a Christmas tree. It is hogging the resources. When you start to sing, the blood has to move to the motor cortex (to move the mouth), the auditory cortex (to hear the pitch), and the linguistic centers (to process the rhyme).

You are literally draining the blood out of the emotional center to fuel the expression center. You are siphoning the fuel tank of the depression to run the engine of the song.

This is why you feel "emptied out" after singing a sad song. You have burned the fuel.

The danger of the modern "positive vibes only" culture is that it denies us this fuel burning. We are told to "look on the bright side." We are told to "manifest abundance." We are told to listen to high-frequency meditation music.

This is spiritual bypassing. It asks us to ignore the garbage in the basement and pretend we live in a penthouse.

But the garbage is still there. If you don't take it out, it smells.

Singing the Blues is taking out the trash. It is ugly work. It smells. It involves admitting that you are broke, lonely, horny, and scared. But once the trash is on the curb, the house is clean.

We need to reclaim the right to sing ugly songs.

We need to reclaim the "Dirge." A dirge is a slow, mournful song for the dead. It is heavy. It drags. It forces you to walk slowly. It aligns your physical movement with your emotional reality.

When we try to rush grief—when we try to "get back to work" three days after a funeral—we are creating a dissonance. The body is slow, but the schedule is fast. This dissonance tears us apart.

Singing a dirge aligns the inside and the outside. It says, "We are moving slow today. We are heavy today."

This alignment is healing. It stops the war between what you feel and what you do.

The physiological markers of grief—the lump in the throat—are literal. The "lump" is the globus sensation. It is caused by the muscles of the throat (the cricopharyngeus) tightening in a spasm of sympathetic arousal. You are trying to swallow, but the throat is trying to stay open to breathe more air for the "fight." The conflict creates the lump.

Singing massages the lump. The vibration of the vocal folds loosens the cricopharyngeus. It relaxes the constriction. It physically dissolves the lump in the throat.

You cannot talk your way around the lump. You have to vibrate it away.

This is why talk therapy sometimes hits a wall. You can analyze your grief for ten years. You can understand exactly why your mother's death traumatized you. You can have all the vocabulary.

But the trauma is stored in the body, not the intellect. It is stored in the tension of the diaphragm, the tightness of the jaw, the constriction of the throat.

To move it, you need a physical lever. The voice is that lever.

There is a story of a woman in a nursing home who hadn't spoken since her husband died. She sat in a chair, staring at the wall. The nurses tried everything. They brought photos. They brought food. Nothing.

A music therapist came in. She didn't ask the woman to talk. She sat down and started singing an old folk song. A song about loss. "Red River Valley."

"From this valley they say you are going..."

The woman didn't move for the first verse. But on the second verse, her mouth opened. A sound came out. It wasn't a word; it was a groan. A rusty, cracked note.

She started to sing. She sang the line, "Do not hasten to bid me adieu."

And then she started to cry. Not a polite cry. A heaving, shaking sob. The dam broke. The silence of two years was shattered by the song.

The song gave her a vehicle for the pain. It gave her a container that was strong enough to hold the weight of her loss. Words were too weak. The melody was strong enough.

This is the function of the "minor key." In Western music, we associate the minor key with sadness. It has a specific interval—the flattened third—that creates a sense of tension, of unresolve.

The brain reacts to the minor third with a sense of "yearning." It pulls at the heart. It mimics the natural prosody of a sad human voice. When we say, "I'm so sad," our voice naturally falls in a minor interval.

Music codifies this biological signal. It turns the "sound of sad" into a system.

When we sing in a minor key, we are resonating with the frequency of human suffering. We are tuning ourselves to the channel of grief.

This sounds depressing. But it is actually connecting. Because everyone—every single person you pass on the street—is tuned to that channel at some point.

If you refuse to sing the minor key, you are cutting yourself off from half of the human experience. You are living in a major-key delusion.

Life is not a major chord. Life is dissonant. Life is full of unresolved tension.

The Blues admits this. The Blues says, "Yes, the rent is late. Yes, the woman left. Yes, my back hurts."

And then it adds the kicker: "But I'm gonna sing about it anyway."

That "anyway" is the victory.

The victory is not in escaping the pain. The victory is in metabolizing the pain into energy.

When you finish singing a sad song, you don't feel sadder. You feel tired. You feel clean. You feel like you have weathered the storm.

You have metabolized the misery. You have turned the lead into gold. You have eaten the grief and survived.

The silence is the enemy. The silence allows the ghosts to grow. The song forces them into the light.

So when the darkness comes—and it will come—do not be quiet. Do not be polite.

Put on the saddest record you own. Open your mouth. And let the river run.

Chapter 12

Singing in Chains (Endurance & Protest)

In 1947, the ethnomusicologist Alan Lomax drove a heavy recording truck into the Mississippi State Penitentiary at Parchman Farm. He was looking for the roots of the Blues. What he found was the sound of survival in its rawest, most industrial form.

He recorded a group of inmates chopping wood. They were lined up in a row, wielding heavy axes. The heat was oppressive. The labor was lethal. If a man swung too early or too late, he risked striking the man next to him. If he swung too slow, the guard—the "captain"—would punish him.

To survive the shift, the men built a machine out of sound.

One man, the leader, would sing a line: "Old Dollar Mamie..."

The group would grunt in unison as the axes hit the wood: "HUH!"

"She's a-long and tall..."

"HUH!"

The sound of the axe striking the log was the downbeat. The vocalization was the upbeat. The melody provided the grid. By locking their motor cortexes to the rhythm of the song, the men synchronized their movements with millisecond precision. They turned a group of suffering individuals into a single, multi-armed engine. The song dictated the pace. It ensured that no one moved too fast, burning out their energy before noon. It ensured that no one moved too slow, attracting the whip.

This is the "Work Song." It is not art. It is an ergonomic tool.

For most of human history, labor was manual and collective. We moved heavy objects by pulling on ropes. We crushed grain by pounding pestles. We harvested crops by swinging scythes. These actions require peak force output at specific intervals.

If ten men pull on a rope at random times, the force is dissipated. The object does not move. If ten men pull on a rope at the exact same instant, the force is multiplied. To achieve that synchronization without a stopwatch, you need a cue.

The sea shanty is the maritime version of this tool. During the Age of Sail, raising a mainyard or weighing an anchor required the coordinated effort of twenty or thirty sailors hauling on a wet, heavy hemp line. The physical resistance was immense.

The shanty man—a specialist often paid more than the other sailors—would scream out a verse. "Way, hay, and up she rises!" On the word "rises," every man hauled. The rhythm of the lyrics matched the biomechanics of the pull. The verse allowed the men to reset their grip and take a breath. The chorus triggered the exertion.

This acoustic technology prevented injury. When you are hauling a line under thousands of pounds of tension, a slip means a lost finger or a broken back. The song keeps the tension constant. It keeps the crew alert. It provides a dopamine hit (via the melody) to counteract the lactic acid building up in the muscles.

But the utility of the work song goes beyond simple coordination. It functions as a "cognitive dissociation" device.

Pain is attention. When you are suffering—when your muscles are burning, your hands are blistered, and the sun is baking your skin—your brain focuses entirely on the misery. Time dilates. Five minutes feels like an hour. The mind traps itself in the immediate sensation of the body.

Singing forces the mind out of the body.

To sing a complex ballad while swinging a hammer, you have to split your consciousness. Part of your brain is handling the motor task. The other part is handling the narrative of the song. You are thinking about "Old Dollar Mamie." You are thinking about the next rhyme. You are navigating the melody.

This splits the attentional resources. You literally have less brain power available to process the pain signals coming from your hands. The song acts as a distraction filter. It allows the laborer to "go away" for a while.

This is why the longest work songs are often narrative ballads with dozens of verses. They are designed to eat time. They are designed to transport the mind out of the prison yard and back to the woman, the whiskey, or the river.

In the context of slavery and imprisonment, the song serves a third, more subversive function: It creates a private communication channel in a public space.

The "Field Holler" was a solo vocalization used by enslaved people working in the cotton fields. It sounded like a mournful, sliding cry. To the overseer on the horse, it sounded like noise, or perhaps a primitive expression of sadness.

To the workers, it was the internet.

Because low-frequency sound travels over long distances, a field holler could carry information across the plantation. By changing the contour of the melody or inserting coded slang, a worker could signal the location of the overseer. "Captain's coming" could be encoded in a hum. "Water boy is here" could be encoded in a specific interval.

This evolved into the sophisticated steganography of the Spirituals.

Steganography is the practice of hiding a message within another message. The Spirituals were ostensibly religious songs. They used the imagery of the Bible—Moses, Pharaoh, the River Jordan, the Chariot. The slave owners encouraged this. They liked seeing their "property" embracing Christianity. They believed that religion made the slaves docile.

They were wrong. The enslaved population weaponized the theology.

"Wade in the Water" was not just a song about baptism. It was a tactical instruction for escaping runaways. It meant: "Get off the trail and into the river. The slave catchers are using bloodhounds. The water will wash away your scent."

"The Gospel Train" was not about going to heaven. It was about the Underground Railroad. "Swing Low, Sweet Chariot" was a signal that a conductor was in the area, ready to smuggle people north.

"Follow the Drinking Gourd" was an astronomical map. The "Drinking Gourd" is the Big Dipper. The song contains explicit navigational data: "The riverbank makes a very good road / The dead trees will show you the way." It told escaping slaves to follow the Tombigbee River north until it met the Tennessee River, tracking the North Star.

If a slave owner found a written map in a cabin, the slave would be whipped or killed. But they could not find a map hidden in a song. The melody provided "plausible deniability." If challenged, the singer could simply say, "I am just praising the Lord."

This is the genius of singing in chains. It takes the one thing the oppressor cannot confiscate—the voice—and turns it into a multi-tool for survival, escape, and resistance.

This dynamic persists in modern incarceration.

In the Soviet Gulag, silence was a weapon of the state. Prisoners were forbidden to speak during work. They were isolated in solitary confinement. The goal was to atomize the individual, to break their will by removing all social reflection.

But the prisoners developed "tapping" codes on the pipes, and they whispered songs in the barracks. The song was proof of life.

Aleksandr Solzhenitsyn, in The Gulag Archipelago, describes how the simple act of humming a melody could restore a sense of humanity in a place designed to strip it away. The song re-established the boundary of the self. The guards controlled the body, the food, and the time. But they could not control the vibration of the larynx.

To sing in a concentration camp is an act of defiance that borders on the biological. It is a refusal to accept the reality of the environment.

In the Nazi camps, music was often used as torture—orchestras of prisoners forced to play happy marches while work details went to their deaths. But in the shadows, there was "secret singing."

The "Moorsoldaten" (Peat Bog Soldiers) song was written by prisoners in the Börgermoor concentration camp in 1933. The lyrics describe the bleakness of the camp: "Up and down the guards are pacing / No one, no one can go through." It is a song of despair.

But the final verse changes. "Winter cannot last forever / We will say to you goodbye."

When the prisoners sang this, they were asserting a future. They were engaging in "chronological rebellion." The camp was designed to be their end. The song insisted it was temporary.

This insistence is critical for psychological survival. Viktor Frankl, the psychiatrist who survived Auschwitz, argued that the prisoners who lost the ability to imagine a future were the ones who died first. The song kept the future alive. It kept the horizon open.

The "Singing Revolution" of the Baltic states in the late 1980s is the ultimate macroscopic example of this power.

Estonia is a small nation. In 1987, it had a population of about 1.5 million. It was occupied by the Soviet Union. The Soviets had a massive military presence—tanks, jets, KGB operatives. The Estonians had no weapons. A violent uprising would have been suicide.

So they used the only infrastructure they had: the choir.

Estonia has a tradition of "Laulupidu"—a massive song festival held every five years. It involves choirs of up to 30,000 singers performing for audiences of hundreds of thousands. It is a sonic event that rivals a nuclear detonation in terms of energy.

In the summer of 1988, the festivals turned into protests. At the Tallinn Song Festival Grounds, a massive concrete amphitheater, over 300,000 Estonians gathered. That is one-third of the entire nation in a single field.

They didn't burn cars. They didn't throw rocks. They held hands and they sang.

They sang the old national anthems that had been banned by the Soviets. They sang folk songs about the land. They sang rock songs written by patriotic composers.

Imagine being a Soviet commander in that situation. You have orders to maintain control. But how do you control 300,000 people who are standing still and singing in perfect harmony?

If they were rioting, you could shoot them. The chaos would justify the force. But the order of the singing creates a "moral force field." The discipline of the choir matches the discipline of the army. It creates a symmetry.

The singing de-legitimized the occupation. It made the tanks look ridiculous. It made the guns look impotent. The sound of the choir was the sound of a coherent, unified culture asserting its right to exist. It was a sonic wall that the Soviet ideology could not breach.

The revolution spread to Latvia and Lithuania. On August 23, 1989, two million people joined hands to form a human chain spanning 400 miles across the three countries. The "Baltic Way." They passed the signal of freedom hand-to-hand, and they sang the whole time.

The Soviet Union collapsed shortly after. The Baltic states regained their independence without a civil war. They sang their way out of the empire.

This historical anomaly proves that the voice is a political instrument. It works because it solves the "collective action problem."

In any revolution, the individual is afraid to step forward. "If I protest alone, I get shot." You need a mechanism that allows everyone to step forward at the exact same time, so there is no single target.

Singing provides that mechanism.

You cannot sing a choral anthem alone. It requires the group. When the song starts, everyone joins in simultaneously. The risk is distributed. The volume provides cover. The individual voice is hidden in the mass, but the mass voice is deafening.

We see this in the Civil Rights Movement in America. "We Shall Overcome" was not just a nice sentiment. It was a tactical tool for non-violent resistance.

When protesters were being hosed by police, or dragged into paddy wagons, they sang. Why?

First, for the breath. As we established, singing prevents the panic response. It keeps the heart rate manageable. It prevents the protesters from breaking into a chaotic run, which would trigger a violent police response.

Second, for the optics. A singing protester looks peaceful. A screaming protester looks dangerous. The song frames the narrative for the cameras. It captures the moral high ground.

Third, for the endurance. A protest march takes hours. Standing in the cold takes hours. The song provides the rhythm to keep walking. It generates the body heat and the emotional heat required to stay on the line.

The song "We Shall Overcome" is structurally simple. It is slow. It is repetitive. It is easy to harmonize. This is intentional. It is "open source" technology. You don't need to be a musician to use it. You just need a throat.

The melody itself is calming. It uses long, sustained notes. It resolves clearly. It is a "major key" promise of victory delivered with a "slow tempo" patience. It tells the nervous system: "We are not in a rush. We are inevitable."

This inevitability is the core message of the protest song. It tells the oppressor: "You can hurt our bodies, but you cannot stop the vibration."

The physics of sound supports this. You cannot arrest a sound wave. You cannot build a wall high enough to block a low-frequency hum. Sound permeates. It leaks through cracks. It travels through walls.

When a prisoner sings in his cell, the sound travels through the ventilation shafts. It reaches the cell next door. It reaches the guard station. It changes the atmosphere of the entire block. It transforms the prison from a collection of isolated cages into a resonant chamber.

This is why silence is the first tool of the tyrant. The first thing a repressive regime does is seize the radio stations. The second thing they do is ban public gatherings. The third thing they do is censor the artists.

They know that the song is the spark.

In the modern era, we have seen this play out in the Arab Spring, in Hong Kong, and in Iran. In Tahrir Square, the chanting was the heartbeat of the revolution. In Hong Kong, the song "Glory to Hong Kong" became the unofficial anthem. The government eventually banned it. They tried to scrub it from the internet.

But you cannot scrub a melody from a brain. Once the song is installed in the population, it stays there. It becomes a sleeper agent, waiting for the next gathering to be activated.

The "Chain" in the title of this chapter refers to the physical shackles, but it also refers to the acoustic chain. The voice connects one person to the next. It links the past to the present.

When you sing a work song today, even if you are just gardening or doing the dishes, you are tapping into that ancestral lineage. You are using the same neuromuscular patterns that built the pyramids and laid the railroads.

You are using the rhythm to trick your brain into enduring the effort.

We have largely lost this in the white-collar world. We work in silence. We type. We click. The only rhythm is the hum of the HVAC system.

This silence makes the work harder. It makes the day feel longer. We have removed the acoustic lubricant that makes labor bearable.

We listen to music while we work—headphones on—but that is consumption, not production. We are letting someone else set the pace.

The true power comes when you generate the rhythm yourself.

This is why military units still run to cadence. This is why sports teams chant. They know that shared suffering is easier than solitary suffering. If we all groan together, the load is lighter.

The lesson from the chain gang and the concentration camp is stark: The voice is the one freedom that cannot be fully stripped.

You can take a man's clothes. You can take his food. You can take his name. You can shave his head. But as long as he has air in his lungs, he can hack the air. He can create a vibration that asserts his existence.

He can say, "I am here."

And if someone else answers—"HUH!"—then he is not alone. And if he is not alone, he can survive.

Chapter 13

The Silence of the Lambs (The Cultural Shift)

On December 6, 1877, Thomas Edison walked into the office of *Scientific American* in New York City. He placed a small machine on the editor's desk. It was a crude device, consisting of a grooved brass cylinder wrapped in tinfoil, a hand crank, and a needle diaphragm. Edison turned the crank.

The machine spoke. "Good morning. How do you do? How do you like the phonograph?"

The editors were terrified. They were men of science, accustomed to steam engines and telegraphs, but this was different. This was a violation of a fundamental law of physics. Until that moment, sound was ephemeral. It was an event. It happened, it vibrated the air, and then it vanished forever. You could not hold it. You could not store it. You certainly could not sell it.

Edison changed sound from an event into an object. By capturing the waveform in tinfoil, he froze time. He made the human voice repeatable.

This invention is usually celebrated as a triumph of communication. We gained the ability to hear the voices of the dead, to listen to symphonies in our living rooms, and to transmit culture globally. But biologically, the phonograph was a disaster. It marked the beginning of the Great Silence.

Before 1877, if you wanted to hear music, you had to make it.

This was a logistical constraint. There was no radio. There was no Spotify. There was no background noise. If you wanted a song at a birthday party, someone in the room had to sing it. If you wanted a hymn at a funeral, the congregation had to generate the volume. If a mother wanted her child to sleep, she had to provide the lullaby herself.

This necessity created a culture of participation. In the Victorian era, the piano was the center of the middle-class home. Sheet music was the primary product of the music industry. People bought the code (the notes) and executed the software (the song) themselves.

Competence was widespread. It wasn't "professional" competence, but it was functional. A typical family could hold a four-part harmony. A typical pub crowd could carry a tune. Singing was a utility, like cooking or sewing. You didn't have to be a Michelin-star chef to cook dinner for your family, and you didn't have to be an opera star to sing for your friends. You just did it because it needed to be done.

The phonograph broke this dynamic. It introduced the concept of the "Reference Recording."

Suddenly, you could hear the best singer in the world in your own living room. You could hear Enrico Caruso. You could hear Bessie Smith. And inevitably, you compared your own voice to theirs.

The comparison was fatal. The professional recording was pitch-perfect. It was consistent. It didn't crack. It didn't get tired. It didn't forget the lyrics. Compared to the polished, mechanical perfection of the record, the amateur voice sounded rough, thin, and embarrassing.

We began to outsource the labor. We stopped buying sheet music and started buying records. We stopped gathering around the piano and started gathering around the Victrola. We shifted from a culture of producers to a culture of consumers.

This shift created a new binary: The Talent versus The Audience.

Before recording, singing was a spectrum. There were great singers, good singers, average singers, and bad singers. Everyone was on the graph. After recording, the graph became a cliff. On one side were the "Artists"—the genetic elites, the professionals, the ones with the contracts. On the other side was everyone else.

We accepted a new social contract: The Artists make the noise; the Audience pays for the noise. The price of admission is silence.

This contract is unique to music. We do not apply it to other physical functions. We watch professional athletes run 100 meters in under ten seconds, but we do not stop jogging. We do not say, "Well, I can't run like Usain Bolt, so I will never move my legs again." We understand that running is a function of the human body that maintains health, regardless of elite performance.

We watch professional chefs on television, but we still make scrambled eggs. We acknowledge that there is a difference between "cooking for nourishment" and "cooking for a Michelin star."

But with singing, we bought the lie. We accepted the idea that unless you are "gifted," you should keep your mouth shut.

This is the "Talent Myth." It is the pervasive cultural belief that singing is a magical trait bestowed upon a lucky few at birth, like blue eyes or double-jointed thumbs. If you have "it," you sing. If you don't have "it," you listen.

Biologically, this is nonsense.

Barring severe pathology or injury, every human being has the same equipment. You have lungs. You have a larynx. You have vocal folds. You have a tongue. You have a brain that can process pitch. The ability to sing is not a binary trait; it is a coordination skill. It is exactly like learning to catch a ball or ride a bicycle.

Some people have a genetic predisposition for faster muscle twitch or better auditory processing, just as some people are naturally faster runners. But the baseline ability to carry a tune is hard-wired into the species. As we established in previous chapters, we are built to do this.

The "tone deaf" statistic is the most damning evidence of this myth. True amusia—the neurological inability to distinguish pitch—affects approximately 1.5% to 4% of the population. If you can tell the difference between the voice of a man and the voice of a woman, or if you can tell when someone asks a question (because the pitch goes up at the end), you are not tone deaf.

Yet, if you survey adults in Western countries, nearly 50% will claim they "cannot sing" or are "tone deaf."

They are not disabled. They are atrophied.

They stopped singing when they were children, usually because of a specific moment of shame. A teacher told them to mouth the words in the choir. A sibling laughed at them. A parent told them to be quiet.

Because they stopped, the neural pathways degraded. The coordination between the ear and the larynx grew rusty. Now, when they try to sing, they really are bad. They struggle to match pitch. Their voice cracks. This confirms the bias: "See, I don't have the gift."

It is a self-fulfilling prophecy. We silenced the children, and then we mocked the adults for being silent.

The recording industry relies on this insecurity. If everyone felt confident singing their own songs, the value of the recorded product would drop. The industry needs you to believe that music is a commodity that must be purchased from authorized vendors. They need you to believe that your own voice is inadequate.

This commodification reached its apex with the rise of reality television.

In 2002, American Idol premiered. It became the most popular show in the history of American television. It turned the "Talent Myth" into a blood sport.

The format of the show was specific. It began with the "audition rounds." These episodes were not designed to find talent; they were designed to expose the lack of it. We watched thousands of deluded amateurs enter a room, sing horribly, and get destroyed by a panel of judges.

Simon Cowell became the avatar of the new cultural superego. He was the voice of "objective standards." He told people they were dreadful. He told them they sounded like dying animals. He told them to go home and never sing again.

The audience laughed. We sat on our couches and jeered. "Look at that idiot," we said. "He thinks he can sing. How embarrassing."

We were participating in a ritual of social enforcement. The show was teaching a lesson: Do not step out of your lane. If you are not "good," you will be punished. You will be humiliated on a national stage.

The message was clear: Singing is an audition. Every time you open your mouth, you are being judged. If you are not world-class, you are a joke.

This terror of judgment has filtered down into every aspect of life. Go to a birthday party. When it comes time to sing "Happy Birthday"—a song with a range of one octave, designed for children—the adults freeze. They mutter. They sing in a low, ironic monotone. They are terrified of being heard.

We have created a culture of vocal anorexia. We are starving ourselves of the vibration we need to survive because we are afraid of looking fat in the mirror of public opinion.

The economic impact of the recording invention is easy to track. The physiological impact is harder to measure, but it is visible in the public health data.

As we stopped singing, our respiratory health declined. Our stress levels skyrocketed. Our social cohesion fractured.

We replaced the active regulation of the nervous system (singing) with the passive consumption of regulation (listening).

When you listen to a sad song on Spotify, you get a shadow of the chemical release. You get a little empathy. You get a little catharsis. But you don't get the vibration. You don't get the deep breath. You don't get the vagal massage.

It is the difference between watching a massage on YouTube and actually getting a massage. One is information; the other is physical manipulation.

We have become a nation of voyeurs. We watch people sing. We watch people dance. We watch people cook. We watch people have sex. We consume the image of the act, but we do not perform the act.

This passivity changes the brain.

The motor cortex—the strip of the brain that controls movement—needs to be used. If you don't use the fine motor control of the larynx, the map in the brain gets fuzzy. The "representation" of the throat shrinks.

This is use-dependent plasticity. A violinist has a massive area of the brain dedicated to the left hand. A singer has a massive area dedicated to the vocal tract. The average modern human has a shriveled, dusty map where their voice should be.

This atrophy affects speech. We are seeing a rise in vocal fatigue, vocal nodules, and weak, breathy speaking voices. We are losing the ability to project. We rely on microphones for everything. Teachers wear microphones in classrooms. Fitness instructors wear microphones. We have forgotten how to use the bellows of the body to fill a room with sound.

We are whispering our way to the grave.

The technology continues to advance, widening the gap.

First, it was the microphone, which allowed singers to croon softly instead of projecting. Then it was reverb, which smoothed out the rough edges. Then it was multi-tracking, which allowed a singer to harmonize with themselves.

Then came Auto-Tune.

Invented in 1997 by Andy Hildebrand, an engineer who worked in seismic data processing for the oil industry, Auto-Tune was designed to correct pitch drift. It takes the input signal, analyzes the frequency, and snaps it to the nearest perfect semitone.

If you sing a little sharp, the computer fixes it. If you sing a little flat, the computer fixes it.

Suddenly, perfection was not just attainable; it was mandatory.

Before Auto-Tune, a recorded vocal had humanity. If you listen to a Beatles record, or an Aretha Franklin record, there are "mistakes." There are notes that are slightly sharp or slightly flat. These imperfections create tension. They create emotion. They sound like a human being pushing air.

After Auto-Tune, the radio became a stream of mathematical perfection. Every note hit the center of the target. The "human element"—the waver, the slide, the struggle—was erased.

This created an impossible standard. We are now comparing our raw, acoustic voices not just to professionals, but to cyborgs. We are competing with robots.

When a teenager sings in the shower and hears their voice waver, they think they are failing. They don't realize that the voice on the radio didn't waver because it was run through a microprocessor. They are comparing their biology to an algorithm.

This drives the silence deeper. Why bother trying if you can't sound like the machine?

The silence has spread to our public spaces.

Walk into a church. Fifty years ago, the congregation sang. The sound was deafening. Today, in many modern churches, the congregation stands in the dark while a professional band performs on a stage with lights and smoke machines. The sound system is so loud that you couldn't hear yourself sing if you tried.

The parishioners are not participants; they are an audience. They are watching a concert. The communal act of worship has been replaced by the consumption of religious entertainment.

Walk into a pub. The "pub sing" is dead in most places. It has been replaced by the jukebox or the DJ. If someone starts singing, they are asked to leave. They are "disturbing the peace."

The only exception is Karaoke.

Karaoke is a fascinating cultural artifact. It is the designated "safe zone" for amateur singing. But even Karaoke is framed as irony. It is usually done drunk. It is done as a joke. You are supposed to be bad. If you are too good at Karaoke, it's awkward. The social contract allows us to sing only if we frame it as a comedy routine. We have to laugh at ourselves before anyone else can.

This preemptive defense mechanism protects the ego, but it neutralizes the power of the act. You cannot get the deep, tribal connection of the "Human Howl" if you are treating the song as a joke. You cannot vibrate the Vagus nerve effectively if you are giggling.

The "Silence of the Lambs" is not just about the loss of music. It is about the loss of agency.

When we stop making our own noise, we become easier to control. A silent population is a compliant population. If we are content to listen to the voices of the wealthy and the famous, we are less likely to raise our own voices in dissent.

We have accepted a hierarchy where a few people speak and everyone else listens. This is dangerous for democracy. It is dangerous for mental health.

The "consumer mindset" treats music as a product you buy to make yourself feel better, like a candy bar. "I'm sad, I'll play some Adele." "I'm happy, I'll play some Pharrell."

But the biological function of music is not consumption; it is production. The benefit comes from the effort.

You cannot buy physical fitness. You have to run the miles. You cannot buy vocal fitness. You have to sing the notes.

We are currently living in the "Dark Ages" of the voice. We have more access to music than any generation in history, yet we are the least musical generation in history. We have traded the birthright of the throat for the convenience of the earbud.

This is a bad trade.

We gave up the oxytocin of the group sing. We gave up the cortisol reduction of the hum. We gave up the dopamine of the rhythm. We gave up the social glue of the harmony.

In exchange, we got high-fidelity recordings of other people having those experiences.

We are starving to death in a supermarket.

The way out of this trap is to break the phonograph. Not literally—you can keep your iPhone—but metaphorically. You have to reject the idea that recorded music is the "real" music and your voice is the "fake" music.

The opposite is true.

Recorded music is a ghost. It is a fossil of a sound that happened in the past. Your voice, right now, cracked and imperfect as it is, is the only living thing in the room.

The amateur voice is the authentic voice. The word "amateur" comes from the Latin amator, meaning "lover." An amateur is someone who does it for love, not for money.

We need more lovers. We need fewer professionals.

We need to stop worrying about being "discovered" and start worrying about being heard. We need to stop waiting for the judges to give us permission.

There are no judges. Simon Cowell is just a man with a high-waisted pant and a business model. He does not own the physics of sound.

The air is free. The vibration is free. The machinery is installed.

The silence is a choice. And every day you don't sing, you are choosing to let the atrophy win. You are choosing to let the machine speak for you.

The phonograph was a miracle, but it was also a mute button. It is time to unmute.

Chapter 14

The Karaoke Paradox (Vulnerability & Shame)

The room is dark. The air smells of stale beer and ozone. On a small stage in the corner, a man in a wrinkled dress shirt is holding a microphone. He is swaying slightly. He is looking at a screen that displays the lyrics to "Don't Stop Believin'" by Journey.

He is terrified.

If you hooked this man up to a heart rate monitor, you would see a spike consistent with a skydiving jump. His palms are sweating. His mouth is dry. His cortisol levels are redlining. He is about to engage in an act that his primitive brain interprets as a suicide mission. He is about to reveal his genetic fitness—or lack thereof—to a room full of potential mates and rivals.

He opens his mouth. He misses the first note by a semitone. He cracks on the high part. He is flat. He is loud. He is objectively terrible.

And the room loves him for it.

This is the Karaoke Paradox. It is the only performance space on earth where failure is not just tolerated; it is celebrated. It is a modern ritual of vulnerability that allows us to bypass the "Talent Myth" and access the primal utility of the voice, provided we are drunk enough to try it.

To understand why karaoke exists, you have to understand the biology of shame.

Shame is a survival mechanism. It is the emotional system that keeps us in line with the tribe. In a prehistoric context, if you violated a social norm—if you hoarded food, or stole a mate, or acted erratically—the tribe would ostracize you. In the wild, exile meant death. Therefore, the brain evolved a powerful, painful alarm system to prevent us from doing anything that might lead to rejection.

That alarm system is the fear of public humiliation.

Surveys consistently rank public speaking as the number one fear of the average adult, often ranking higher than death. People would rather be in the casket than giving the eulogy. This makes sense. Death is abstract. Rejection is immediate.

Public singing is public speaking on steroids. When you speak, you are transmitting information. When you sing, you are transmitting emotion. You are sustaining tones. You are exposing the physical capability of your respiratory system. You are opening the "kimono" of your biology.

If you sing badly, you are broadcasting a signal of incompetence. You are saying, "My motor control is poor. My hearing is bad. I am not a high-value specimen."

The shame response to this is visceral. The capillaries in the face dilate (blushing). The head drops. The shoulders roll forward. The body tries to make itself small. It tries to disappear.

This is why the karaoke bar is dark. The darkness provides a layer of anonymity. It lowers the stakes.

But the darkness isn't enough. We need chemistry.

It is no accident that karaoke is inextricably linked to alcohol. You rarely see a dry karaoke bar. The function of the alcohol is not just to lower inhibitions; it is to specifically anesthetize the prefrontal cortex.

The prefrontal cortex is the CEO of the brain. It handles judgment, planning, and impulse control. It is the voice in your head that says, "Don't do that, you'll look stupid." It is the center of the Talent Myth.

When you ingest ethanol, the prefrontal cortex goes offline. The "internal critic" is sedated. The connection between the embarrassment center and the motor center is severed.

This allows the "howl" to escape. Under the influence of three margaritas, the biological imperative to make noise finally overpowers the social imperative to be quiet. The gatekeeper is asleep, so the animal grabs the mic.

This creates a unique social environment. A karaoke bar is a "safe space" in the most literal, neurological sense. It is a Temporary Autonomous Zone where the rules of social hierarchy are suspended.

In the office, the man in the wrinkled shirt is a junior accountant. He is low status. He has to be quiet. In the karaoke bar, for three minutes, he is the alpha. He holds the totem (the microphone). He commands the attention of the tribe.

The audience plays a crucial role in this contract. In a professional concert, the audience is critical. They paid $100 for a ticket. They expect perfection. If the singer fails, the audience feels ripped off. They boo.

In a karaoke bar, the audience is complicit. They didn't pay for excellence; they paid for participation. They know that in ten minutes, it will be their turn.

This creates a dynamic of "reciprocal vulnerability." If I cheer for you when you suck, you will cheer for me when I suck. We are signing a non-aggression pact.

This is why the "good" singer is often the most hated person in the karaoke bar.

We all know this person. They bring their own microphone. They don't drink. They pick a difficult ballad—maybe Whitney Houston or Celine Dion. They stand center stage. They close their eyes. They deliver a technically flawless, pitch-perfect performance.

The room goes cold. People stop talking. They clap politely at the end, but the energy is weird. It feels hostile.

Why? Because the "good" singer violated the contract. They were not being vulnerable; they were showing off. They were peacocking. They raised the standard of the room back up to "professional," which makes everyone else feel unsafe. They reminded the tribe of the hierarchy.

The hero of the karaoke bar is the person who picks "Bohemian Rhapsody," cannot hit the high notes, forgets the words, falls off the stage, and keeps singing.

We love that person because they are proving that the fear is a liar. They are surviving the social death. They are looking foolish, and the roof is not caving in. The tribe is not exiling them. In fact, the tribe is singing along.

This sing-along moment is the peak of the ritual. When the singer falters, the room steps in. They provide the "backing vocals." This is an act of rescue. It is a sonic safety net. It says, "You are losing the melody, so we will hold it for you."

The physiological effect of this rescue is massive. The singer floods with oxytocin. The fear of rejection is transmuted into the joy of acceptance. They walk off the stage feeling high. They have faced the dragon and lived.

The Japanese invented karaoke in the 1970s ("kara" means empty, "oke" means orchestra). It was originally designed for businessmen. In Japanese corporate culture, hierarchy is rigid. You do not speak out of turn. You do not show emotion. You do not question the boss.

Karaoke provided the release valve. After work, the boss and the subordinates would go to a box, get drunk, and sing. The junior employee could sing a heavy metal song and scream in the boss's face. The boss could sing a sentimental love song and cry.

It was a sanctioned method of emotional dumping. It allowed the men to bond without breaking the formal rules of the office. "What happens in the karaoke box stays in the karaoke box."

This spread to the West because we are equally repressed, just in different ways. We are repressed by the demand for "cool." We are terrified of being "cringe."

Karaoke is the weaponization of cringe. It forces you to be uncool. You cannot look cool singing "Sweet Caroline." It is impossible. You have to lean into the cheese. You have to embrace the cliché.

This embrace is healing. It breaks the armor of irony that most modern adults wear. You cannot be ironic when you are belting a power ballad. The physical exertion requires sincerity. You have to commit to the bit.

This commitment is what we are starving for. We live in a world of guarded interactions. We text instead of calling. We curate our Instagram feeds to look perfect. We hide our flaws.

Karaoke is the anti-Instagram. It is live, unedited, and ugly. It is the only place where we see our friends actually struggle.

There is a diagnostic utility to the song choice. The song a person picks tells you more about their internal state than a year of therapy.

The quiet girl who picks heavy metal is revealing a reservoir of suppressed rage. The tough guy who picks a Disney princess song is revealing a need for softness. The person who picks a seven-minute song with no chorus is revealing that they are a narcissist who lacks social awareness.

The "Karaoke Paradox" also highlights the importance of the screen.

In a traditional sing-along, you have to know the words. This is a barrier to entry. If you don't know the verse, you can't participate.

The screen removes the memory requirement. It provides the "teleprompter." This democratizes the act. You don't need to be a fan of the band; you just need to be able to read.

But the screen also acts as a focal point. It diverts the gaze.

When someone is singing on a stage, normally the audience looks at the singer. This "gaze pressure" is intense. It feels like judgment.

In karaoke, the audience often looks at the screen, not the singer. They are reading the lyrics along with them. This triangulates the attention. The singer is not the sole target. The song is the target. The singer is just the vehicle.

This reduces the social anxiety. You are not performing at the people; you are performing with the machine, and the people are witnessing it.

The lyrics on the screen are often color-coded, changing as the music plays. This is a visual guide to rhythm. It forces the singer to sync their visual processing with their vocal motor control. It is a high-cognitive-load task.

This explains why people often look so serious while singing a fun song. They are concentrating. They are processing data. They are trying to keep the arrow in the green zone.

When they nail it—when the color change matches the voice perfectly—there is a dopamine hit of "task completion." It feels like a video game.

But unlike a video game, the physical output is vibration.

The sonic environment of the karaoke bar is usually terrible. The speakers are blown. The acoustics are reflective. The microphone smells like saliva and gin.

This lo-fi quality is a feature, not a bug. If the sound system were pristine, the flaws in the voice would be more obvious. The mud hides the mistakes. The reverb (the "echo" knob) is the amateur's best friend. It blurs the pitch. It makes a thin voice sound big.

Karaoke machines usually have the reverb cranked to the maximum. This is acoustic makeup. It allows the singer to feel, for a moment, like they are in a stadium.

The "Karaoke Paradox" is ultimately about permission.

As children, we give ourselves permission to play. We make up songs. We dance. We don't care who is watching.

As we hit puberty, the "imaginary audience" appears. We become hyper-aware of social rank. We stop playing. We start performing.

Karaoke gives us a permission slip to regress. It says, "For the next four minutes, the rules of puberty are suspended. You can be a child again. You can make noise just for the sake of making noise."

This regression is necessary. We carry a heavy load of adult responsibility. We have to be competent at our jobs. We have to be competent parents. We have to be competent partners.

Karaoke is the one place where we are allowed to be incompetent.

You can't be an incompetent surgeon. You can't be an incompetent driver. But you can be an incompetent singer, and the world keeps turning.

Experiencing this lack of consequence is liberating. It teaches the brain that failure is not fatal.

If you can stand up in front of fifty strangers, crack on a high note, have everyone laugh, and then sit down and drink a beer, you have rewired your fear response. You have inoculated yourself against shame.

This confidence bleeds over into the rest of life. If you can survive "I Will Always Love You" in a dive bar, you can survive a presentation in the boardroom. You have hardened the shell.

The "Karaoke Paradox" also reveals our deep need for the "canon."

Karaoke songs are a specific subset of music. Not every song works. Obscure indie rock doesn't work. Jazz doesn't work.

The songs that work are the "Anthems." "Sweet Caroline." "Livin' on a Prayer." "Bohemian Rhapsody." "Piano Man."

These songs share common DNA. They have simple, soaring melodies. They have shouted choruses. They are emotionally broad. They are the secular hymnal of the Western world.

We need these hymns. We need a shared repertoire that everyone in the room knows. When the opening piano riff of "Don't Stop Believin'" starts, there is a collective recognition. The room unifies. We are all members of the Church of Journey.

This shared knowledge allows for the "call and response." The singer sings "Just a small town girl," and the room screams "LIVING IN A LONELY WORLD."

This is the exact same mechanic as a liturgical church service. The priest chants, the congregation responds. It confirms our membership in the group. It affirms that we share a language.

In a fragmented culture, these pop songs are some of the last things we actually agree on. We might fight about politics, but we agree that the "Ba-Ba-Ba" in "Sweet Caroline" must be shouted at maximum volume.

The karaoke bar is the parish of the secular age.

There is a specific phenomenon known as "Karaoke Bonding." You can know a coworker for five years and never really connect. You talk about spreadsheets. You talk about the weather.

Then you go to karaoke. You see them take off their tie and rap "Lose Yourself" by Eminem with intense, terrifying precision.

Suddenly, you know them. You have seen their shadow side. You have seen their passion. You have shared a visceral experience. The relationship changes. The professional distance collapses.

This is why companies in Asia mandate karaoke. They know it fast-tracks trust.

It is harder to screw over someone you have sung a duet with. The harmony creates a bond.

The "Duet" is the highest form of karaoke risk. You are relying on another person. You have to listen to them. You have to blend.

"Summer Nights" from Grease is a classic. It requires a male part and a female part (or at least two distinct energies). It requires acting. It requires looking at each other.

When two people pull off a duet, they create a "third entity"—the sound of the pair. It is a public display of cooperation. Even if they hate each other, for those three minutes, they are partners.

The "Karaoke Paradox" teaches us that we are desperate to be heard, but we are terrified of listening to ourselves.

The sound of your own voice on a PA system is alienating. It sounds different than it does in your head. In your head, your voice is conducted through bone. It is deeper. Richer.

On the speakers, it is air-conducted. It sounds higher. Thinner. This is why people cringe when they hear a recording of themselves. "Do I really sound like that?"

Yes, you do.

Karaoke forces you to confront that reality. You are hearing yourself as the world hears you. It is a reality check.

Accepting that sound—accepting that you are a high-pitched, nasal, imperfect human being—is a form of radical self-acceptance.

You are saying, "This is my instrument. It is not a Stradivarius. It is a kazoo. But I am going to play it with everything I have."

There is dignity in the kazoo.

We need to protect the karaoke bar. Gentrification is threatening it. We are seeing "luxury karaoke" with private suites and bottle service. This kills the dynamic.

The magic of karaoke requires the public room. It requires the strangers. If you are only singing in a private box with your best friends, you are not risking enough. You are not facing the tribe.

You need the drunk guy at the bar who doesn't know you. You need the risk of the heckler. You need the danger.

Without the danger, there is no courage. And without courage, there is no release.

The next time you are in a bar and you see the machine, don't hide. Don't check your phone.

Go to the book. Find the song you loved when you were seventeen. Find the song that makes you cry in the car.

Put your name on the list.

Wait for the call.

When you walk up to that stage, your heart will hammer. Your throat will tighten. Your brain will scream "Abort."

Ignore the brain. Grab the mic. Look at the screen.
And when the words turn green, howl.
You will be flat. You will be off-beat. You will look like an idiot.
And it will be the most honest thing you have done all year.

Chapter 15

The Robot's Lullaby (The Future of Voice)

In 1997, Dr. Andy Hildebrand was working for Exxon. He was a geophysicist specializing in the analysis of seismic data. His job was to send sound waves into the earth and interpret the reflections to find oil deposits. The math required to do this—autocorrelation—was complex. It involved cleaning up noisy signals to find the true structural lines of the bedrock.

Hildebrand left the oil industry and realized the same math could be applied to the human voice. The voice is a noisy signal. It wavers. It drifts. It misses the target frequency. Hildebrand wrote a piece of software that could detect the intended pitch of a singer and instantly correct it to the nearest semitone. He called it Auto-Tune.

He intended it to be a studio tool, a way to fix a stray note in an otherwise good take. He did not intend to rewrite the definition of music.

But in 1998, the singer Cher released "Believe." In the chorus, her producer cranked the "retune speed" to zero. Instead of gently sliding the pitch to the correct note, the software snapped it instantly. The result was a robotic, stair-step warble. It sounded like a cyborg was singing. It was a massive hit.

Since then, the software has moved from a special effect to a standard operating procedure. Today, almost every song you hear on the radio has been pitch-corrected. Even the "natural" sounding vocals are run through Melodyne or Auto-Tune to ensure they sit perfectly on the grid.

We have successfully removed the error from the human voice. In doing so, we have removed the humanity.

To understand why this matters, you have to look at the physics of emotion. When a human being is emotional—when they are crying, angry, or terrified—the laryngeal muscles tremor. The vocal folds swell. The breath becomes erratic. This causes the pitch to fluctuate. It creates "micro-tones" that exist between the keys of the piano.

These fluctuations are the biological signature of authenticity. When you hear a singer's voice crack or slide slightly flat during a sad lyric, your mirror neurons fire. You recognize the sound of distress. You feel it.

Auto-Tune erases this tremor. It takes the messy, jagged waveform of grief and flattens it into a perfect sine wave. It quantizes the emotion.

The result is "Ear Candy." It is sweet. It is shiny. It is mathematically perfect. But it is nutritionally void. It slides into the ear without resistance, and it leaves the listener feeling nothing. We tap our feet, but we do not weep.

We are currently raising a generation of children who have never heard a raw, uncorrected human voice on a recording. Their baseline for "normal" is a mathematical impossibility. They are trying to emulate a machine.

This is the "Robot's Lullaby." It is the sound of a species outsourcing its own vibration to an algorithm.

The problem has escalated beyond pitch correction. We are now in the era of synthesis.

Artificial Intelligence can now generate vocals that are indistinguishable from human singers. In 2023, a track titled "Heart on My Sleeve" went viral. It featured the voices of Drake and The Weeknd. It was catchy. It was well-produced.

But neither Drake nor The Weeknd had anything to do with it. A user named "Ghostwriter977" had trained an AI model on their back catalogs and generated the audio.

The legal ramifications are messy, but the biological ramifications are terrifying. We have decoupled the voice from the body.

For 200,000 years, a voice was proof of presence. If you heard a voice, there was a human being nearby. The sound was physically produced by air rushing through a larynx. The "grain" of the voice—the specific rasp and timbre—was the fingerprint of that specific anatomy.

Now, the voice is just a font. It is a skin you can put on any data stream. You can type a sentence and have it read by Frank Sinatra. You can sing a melody into your phone and have it played back by Freddie Mercury.

This creates a crisis of trust. The "Uncanny Valley" is the feeling of revulsion we get when we see a robot that looks almost human but isn't quite right. We are about to enter the "Uncanny Canyon" of audio.

We will listen to a song and we will not know if the singer has lungs. We will not know if they have a heart. We will not know if the emotion we are hearing is the result of a lived experience or a predictive text generator.

This matters because singing is a transfer of energy. As we established in Chapter 8, the "Hive Mind" relies on the synchronization of biological systems. When I sing with you, my heart rate syncs with yours.

You cannot sync your heart rate with a server farm.

The machine does not breathe. It does not have a vagus nerve. It does not have cortisol. It is simulating the output without doing the work of the input.

If we accept this simulation as a substitute, we are in trouble. We are like lab rats pressing a lever for a synthetic dopamine hit, ignoring the real food in the corner.

The danger is that we will stop trying.

Singing is hard. It takes practice. It takes courage. It involves the risk of failure. If an AI can generate a perfect song in three seconds, why should a human spend ten years learning to sing?

This is the "Calculator Argument." We stopped doing mental math when calculators became ubiquitous. We stopped remembering phone numbers when contact lists appeared. Will we stop singing when the bots take over?

There is a very real possibility that singing will become a "legacy skill," practiced only by hobbyists and historians, like weaving or calligraphy. The vast majority of the population will simply consume the generated content. They will be passive receptors of a perfectly optimized algorithmic feed.

This feed will be addictive. The AI will know exactly what melody triggers your dopamine. It will know exactly what tempo keeps you engaged. It will generate music that is "hyper-palatable," like Doritos for the ears.

But Doritos do not build muscle. And algorithmic music does not build soul.

However, there is a counter-argument. There is a possibility that the perfection of the machine will trigger a massive backlash.

We are already seeing the cracks. The resurgence of vinyl records is not just hipster nostalgia. It is a hunger for friction. A record has surface noise. It pops. It crackles. It requires physical handling. It is imperfect.

People are craving the "Live" experience more than ever. Concert tickets are selling at record prices. Why? Because you can't deepfake a stadium. You can't simulate the sweat of the drummer or the mistake the guitarist makes in the solo.

When a human being stands on a stage and screams, there is a transfer of risk. The audience knows that the singer could fail. They could lose their voice. They could fall. That tension makes the performance real.

A robot cannot fail. Therefore, a robot cannot be brave.

This suggests that the future of the human voice lies in its ugliness.

As the machines capture the high ground of perfection—perfect pitch, perfect tone, perfect rhythm—the humans will retreat to the low ground of the glitch. We will value the rasp. We will value the break. We will value the breath.

We saw this happen with photography. When digital cameras became perfect, people started using filters to make their photos look like grainy, scratched Polaroids. We added the noise back in because the noise felt like memory.

We will do the same with voice. The most valuable singers of the future will not be the ones with the prettiest voices. They will be the ones with the most damaged voices.

Think of Johnny Cash's cover of "Hurt." It was recorded when he was 71 years old. His voice is a wreck. It trembles. It sounds like dry leaves dragging on concrete. He is out of breath.

If you ran that vocal through Auto-Tune, you would destroy it. The power of the song is the sound of a dying man confronting his own decay. The "failure" of the voice is the text.

An AI could analyze the waveform of "Hurt" and replicate the timbre of Cash's voice. It could generate a new song that sounds like him. But it would be a lie. Because the AI isn't dying. The AI doesn't have regrets.

The listener can feel the difference. We have a "bullshit detector" located deep in the limbic system. It is evolutionarily old. It knows when a signal is fake.

This detector is our only defense against the robot. We have to train it. We have to learn to listen for the breath. We have to learn to value the struggle.

This shifts the definition of "good singing."

For the last century, "good singing" meant "accurate singing." It meant hitting the notes on the page.

In the age of AI, "good singing" will mean "honest singing." It will mean singing that reveals the biological reality of the singer, flaws and all.

This is a liberation. It frees us from the tyranny of the American Idol standard. If the robots are better at being perfect than we are, we can stop trying to be perfect. We can start being human again.

The punk rock movement of the 1970s was a reaction to the polished, over-produced prog rock of the era. The Sex Pistols couldn't play their instruments. Johnny Rotten couldn't sing. That was the point. They were screaming against the machine.

We are due for a new vocal punk rock. A movement of people who refuse to be pitch-corrected. A movement of people who sing loudly, badly, and aggressively to prove they are not code.

This rebellion will happen in the living room. It will happen in the car. It will happen in the shower.

Every time you sing to your child, you are fighting the machine.

An AI can generate a lullaby. It can play it through a high-fidelity speaker. It can make it sound like a choir of angels.

But the AI cannot monitor the child's breathing in real-time and adjust the tempo to match. The AI cannot feel the child's body relax and lower the volume. The AI cannot transmit the bacterial biome of the parent to the child through the breath.

The "Umbilical Chord" we discussed in Chapter 3 is biological. It relies on the feedback loop between two living nervous systems. A machine can mimic the sound, but it cannot close the loop.

This is the hard limit of technology. It can simulate the output, but it cannot replicate the connection.

The future of voice is a choice between two paths.

Path A is the "Wall-E" scenario. We sit in chairs. We consume content. We let the algorithms soothe us. We stop singing because the machines do it better. Our vocal cords atrophy. Our vagus nerves go dormant. We become silent, lonely bio-units plugged into a digital drip.

Path B is the "Human Howl" scenario. We accept that the machines are better at math, and we let them have it. We let them handle the spreadsheets and the logistics. But we refuse to give them the song.

We reclaim the messy, wet, chaotic act of making noise. We sing at parties. We sing at funerals. We sing in the streets. We prize the crack in the voice. We prize the wrong note.

We treat the voice not as a product to be polished, but as a proof of life.

This proof is necessary because the world is getting stranger. As deepfakes proliferate, we will no longer be able to trust our eyes. We will see videos of politicians saying things they never said. We will see videos of our loved ones doing things they never did.

Reality is becoming malleable.

In a world of digital illusions, the acoustic vibration of a human body in a room is one of the last anchors of truth. If I am standing next to you, and I scream, you feel it. The air moves. The floor shakes. You cannot fake the physics of displacement.

The "unplugged" moment will become the highest luxury. The most expensive concert of 2050 might be a small room with no microphones, no speakers, and no screens. Just a human being pushing air.

We will pay a premium for the inefficiency. We will pay to hear the struggle.

This brings us back to the evolutionary bet.

Nature bet that the voice was worth the risk of choking. We are now facing a new risk: the risk of obsolescence.

If we stop using the voice, we effectively admit that we are obsolete. We admit that we are just outdated hardware running legacy software, waiting to be replaced by a more efficient cloud-based solution.

But if we keep singing, we assert our relevance. We assert that there is something in the biological experience—some chemical, spiritual, visceral residue—that cannot be coded.

We are asserting that the "glitch" is the point.

The robot cannot howl. It can play a recording of a howl. It can synthesize a waveform that looks like a howl. But it cannot feel the terror or the joy that necessitates the howl.

The howl is a response to the condition of being alive and knowing you will die. It is a response to the pain of the dropped larynx and the joy of the oxytocin. It is a meat-based phenomenon.

So, the prediction is this:

The middle will drop out. The "competent but polished" singers will be replaced by AI. The background music, the jingles, the radio pop—that will all go to the bots.

But the raw edges will remain. The screamers, the wailers, the whisperers. The people who sound like dirt and blood. They will become more important, not less.

We will cherish the flaw because the flaw is the signature.

We are standing at the edge of the uncanny canyon. The robots are beckoning us to jump in. They promise us perfect pitch. They promise us eternal youth. They promise us a voice that never tires and never breaks.

Don't jump.

Stay on the cliff. Stay in the wind. Stay in the body.

Let your voice crack. Let it shake. Let it be ugly.

The robot can sing you a lullaby, but it cannot dream. You are the dreamer. You are the animal. You are the glitch.

And the glitch is the only thing worth listening to.

Conclusion

Unmute Yourself

You have reached the end of the argument.

We have spent fifteen chapters dismantling the machinery of the human voice. We have looked at the dropped larynx, that dangerous evolutionary slip that gave us the ability to sing at the risk of choking. We have looked at the vagus nerve, the internal wiring that allows a hum to slow a racing heart. We have looked at the oxytocin, the dopamine, and the endorphins—the chemical rewards the brain dispenses to those who coordinate their breathing with others.

We have looked at the history of the species, from the first hominids humming to their infants on the savannah to the chain gangs swinging hammers in the Delta. We have looked at the war cries, the lullabies, the shanties, and the dirges.

The evidence is consistent. The human voice is not a luxury item. It is not a talent reserved for the genetic elite. It is not a product to be sold by a record label.

It is a survival tool. It is a piece of biological technology that you carry in your throat every second of your life. It is the primary interface for regulating your own nervous system and connecting to the nervous systems of others.

And for most of your life, you have kept it on mute.

This is not an accusation; it is a diagnosis. You live in a culture that has systematically silenced you. You were told in grade school that you couldn't carry a tune. You were told by television that singing was an audition for fame. You were told by the market that music is something you stream, not something you make.

You were trained to be a listener. You were trained to be quiet.

The result of this silence is a population that is anxious, isolated, and chemically imbalanced. We are suffering from a deficiency of vibration. We are trapping our stress in our tissues because we refuse to open the valve and let it out.

The solution is simple, but it is not easy. You have to unmute yourself.

This does not mean you have to quit your job and join a rock band. It does not mean you have to take voice lessons. It does not mean you have to perform. In fact, "performance" is the enemy. Performance implies an audience. Performance implies judgment.

Unmuting is functional. It is about using the voice to do a job.

The first job is regulation.

Tomorrow morning, you will likely encounter stress. You will hit traffic. You will receive an aggressive email. You will worry about money. Your body will react by dumping cortisol into your blood. Your shoulders will tense. Your breath will shorten.

The standard response is to clench. You bite your jaw. You hold your breath. You suffer in silence.

The unmuted response is to hum.

You are in the car. The windows are up. No one can hear you. Open your mouth. Drop your jaw. Make a low, steady sound. Let the air rattle in your chest.

Do not worry about the pitch. The pitch is irrelevant. Focus on the physical sensation of the vibration. Feel it in your sternum. Feel it in your teeth.

You are manually stimulating the vagus nerve. You are telling your parasympathetic nervous system to engage. You are hacking the loop.

It will feel ridiculous at first. You will feel self-conscious, even when you are alone. That is the conditioning talking. That is the voice of the critic. Ignore it. Keep humming until you feel your heart rate drop. It is a mechanical process. If you push the button, the machine works.

The second job is connection.

You have relationships that are starving for signal. You text your partner. You email your parents. You talk to your children. But text is data. It is cold. It lacks the biological signature of your presence.

Sing to them.

This is the most terrifying instruction in the book. It feels intimate to the point of discomfort. But the discomfort is the point.

Sing a lullaby to your child. It does not matter if you sound like a frog. The child does not care about pitch; the child cares about presence. They need to feel the rumble of your chest against their back. They need to hear the familiar distortion of your specific vocal cords. That sound is their safety map. If you outsource that job to a mobile phone playing a recording, you are severing the tether.

Sing "Happy Birthday" at the party. Do not mumble. Do not sing it ironically. Sing it loud. Sing it like you mean it. You are leading a ritual. You are helping a room full of people synchronize their breathing for thirty seconds. If you sing quietly, the room stays fragmented. If you sing loudly, the room coalesces. Be the nucleus.

Sing with your friends. Go to the karaoke bar. Do not sit in the back. Do not look at your phone. Stand up. Pick a song that everyone knows. And scream.

Invite the failure. Let your voice crack. Let the rhythm slip. Laugh about it. The laughter is part of the medicine. It releases the tension. It bonds the group. You are proving to each other that you can survive a mistake. You are proving that your value to the tribe is not based on your perfection, but on your participation.

The third job is resistance.

You live in a noisy world. You are bombarded by sonic weapons—sirens, construction, advertising jingles, the relentless optimized pop of the grocery store. This noise is an invasion. It colonizes your headspace. It disrupts your thought process.

If you are silent, you are a victim of the environment. You are taking the hits.

To unmute is to push back. When you walk down the street and hum, you are creating a force field. You are occupying your own auditory cortex with your own signal. You are declaring that your internal rhythm is stronger than the external chaos.

This is why people whistle in the dark. It is a territorial claim. It says, "I am here. I am not afraid."

We need more whistlers. We need more hummers. We need more people who are willing to disturb the peace of the silence, because the silence is not peaceful. The silence is numb.

There will be pushback.

When you start unmuting, people will look at you. They will ask if you are okay. They will tell you to be quiet.

They are reacting to the breach of the social contract. We have agreed, as a society, to be voiceless. When you break that agreement, it makes people nervous. It reminds them that they are also muted. It triggers their own shame.

Do not let their shame silence you.

The "Talent Myth" is a cage. It is a lie designed to keep you small. It suggests that only the special are allowed to speak.

You are special because you are alive. You are special because you have a larynx that descended over two million years of evolution specifically so you could make this noise. The hardware is the permission slip.

If you have lungs, you are a singer. If you have a heart, you are a drummer. The rest is just practice.

We are heading into a future where the human voice is becoming endangered. The robots are coming. Artificial Intelligence can already generate vocals that are smoother, more accurate, and more pleasing than anything you can produce.

The temptation will be to surrender. Why sing a lullaby when the smart speaker can play a perfect one? Why sing at a party when the playlist is optimized by an algorithm?

We will be sold the idea that human singing is inefficient. We will be told it is obsolete.

But efficiency is not the goal of a human life. Connection is the goal.

The robot cannot connect. It can simulate, but it cannot share. It cannot feel the relief of the exhale. It cannot feel the vibration of the harmony. It is a ghost in the machine.

You are the machine. You are the wet, messy, breathing reality.

Every time you sing, you are voting for humanity. You are voting for the analog over the digital. You are voting for the flaw over the perfection.

This is the ultimate rebellion. In a world of deepfakes and synthetic media, the sound of a raw, unpolished human voice is the only thing that cannot be faked. It is the badge of authenticity.

So, make the noise.

Make it in the shower, where the acoustics are forgiving and the water washes away the judgment.

Make it in the car, where the glass box contains the scream and the motion gives you courage.

Make it in the kitchen, while you chop vegetables, adding a rhythm to the chore.

Make it in the dark, when the grief comes, using the low moan to move the pain out of your chest.

Make it in the light, when the joy comes, using the shout to share the energy with the sky.

Do not wait until you are good. You will never be good enough for the critics, but you are already good enough for the biology. The vagus nerve does not grade you on pitch. The oxytocin does not grade you on tone. The body responds to the effort, not the aesthetic.

The "Human Howl" is not a metaphor. It is a physical fact.

Imagine the first human who ever sang. They were likely cold. They were likely hungry. They were standing in the vast, terrifying darkness of the Pleistocene night, surrounded by predators that could see in the dark and teeth that could crush bone.

They were small. They were slow. They were weak.

But they opened their mouths. They filled their lungs with the air of the hostile world. And they pushed it back out.

They made a sound that said, "I am here."

And then, from the darkness, another voice answered. "I am here too."

And then another. And another.

The voices locked together. They formed a chord. And in that chord, they found a power that no single individual possessed. They found the power to endure the night.

That power is still there. It is waiting in your throat. It has been waiting for you to stop worrying and start vibrating.

The night is still dark. The predators are still out there, even if they look like anxieties and algorithms now.

But you are not alone. You have the chord.

Unmute yourself.

And howl.

APPENDIX

THE SOURCE CODE

This book was built on the shoulders of giants, neuroscientists, ethnomusicologists, and inmates. If you want to check the math, dig deeper into the wiring, or just hear the sounds described in these pages, this is your map.

SECTION I: THE HARDWARE (Anatomy & Neuroscience)

Porges, Stephen W. The Polyvagal Theory: Neurophysiological Foundations of Emotions, Attachment, Communication, and Self-Regulation. W. W. Norton & Company, 2011.

This is the heavy lifting regarding the Vagus nerve (Chapter 9). Porges is the mechanic who figured out how the throat connects to the heart. It is dense, academic, and absolutely essential if you want to understand the "Social Nervous System."

Levitin, Daniel J. This Is Your Brain on Music: The Science of a Human Obsession. Dutton, 2006.

Levitin breaks down the neurochemistry of the "drug deal" in your head (Chapter 4). He explains the dopamine reward system better than anyone else.

Sacks, Oliver. Musicophilia: Tales of Music and the Brain. Knopf, 2007.

The definitive look at how music persists when the brain fails (Chapter 5 and 6). Sacks' case studies on Alzheimer's and aphasia provide the clinical receipts for the "Right Brain Rebellion."

Lieberman, Philip. The Biology and Evolution of Language. Harvard University Press, 1984.

The core arguments regarding the dropped larynx and the risk of choking (Chapter 1). Lieberman treats the vocal tract as an evolutionary trade-off.

Vickhoff, Björn, et al. "Music structure determines heart rate variability of singers." Frontiers in Psychology, 2013.

The "Choir Effect" study (Chapter 8). This is the hard data showing how heart rates synchronize during group singing.

SECTION II: THE SOFTWARE (History & Anthropology)

Lomax, Alan. The Land Where the Blues Began. Pantheon, 1993.

Lomax is the reason we know what the chain gangs sounded like (Chapter 12). He recorded the work songs at Parchman Farm. This book is a brutal, beautiful look at how music functions as a survival tool in hell.

Chatwin, Bruce. The Songlines. Jonathan Cape, 1987.

The primary text for understanding the Aboriginal navigation system (Chapter 6). While some of Chatwin's anthropology is debated, his description of the "singing map" remains the most vivid introduction to the concept.

Ehrenreich, Barbara. Dancing in the Streets: A History of Collective Joy. Metropolitan Books, 2006.

A critical look at how the "festive" (group singing/dancing) was suppressed by church and state, and why we need it back. It parallels the argument in Chapter 13 regarding the silence of the modern world.

Darwin, Charles. The Descent of Man, and Selection in Relation to Sex. John Murray, 1871.

The origin of the "Peacocking" theory (Chapter 2). Darwin's assertion that "musical notes and rhythms were first acquired by the male or female progenitors of mankind for the sake of charming the opposite sex" is the bedrock of the sexual selection argument.

Mithen, Steven. The Singing Neanderthals: The Origins of Music, Language, Mind, and Body. Harvard University Press, 2006.

A deep dive into the "Hmmmmm" communication system of early hominids (Chapter 3). Mithen argues convincingly that we sang before we spoke.

SECTION III: THE TOOLKIT (Practical Applications)

Stephens, Richard, et al. "Swearing as a response to pain." NeuroReport, 2009.

The ice-water bucket study (Chapter 10). Proof that dropping the F-bomb is an analgesic.

Turkle, Sherry. Reclaiming Conversation: The Power of Talk in a Digital Age. Penguin Press, 2015.

While focused on speech, Turkle's work on the isolation of the "screen age" (Chapter 15) supports the thesis that we are losing the biological signal of presence.

Goodman, Steve. Sonic Warfare: Sound, Affect, and the Ecology of Fear. MIT Press, 2010.

A breakdown of how sound is weaponized (Chapter 7), from the LRAD cannon to the use of high-frequency tones in crowd control.

SECTION IV: THE PLAYLIST (Evidence You Can Hear)

Reading about the howl is one thing; hearing it is another. These recordings demonstrate the raw utility of the voice discussed in the book.

1. "Po' Lazarus" – James Carter and the Prisoners (Recorded by Alan Lomax, 1959)

Context: Chapter 12. Listen to the axes falling. Listen to the grunt. This is the sound of synchronized labor.

2. "Haka (Ka Mate)" – The All Blacks

Context: Chapter 7. Watch a video of this. Observe the entrainment of the team and the intimidation factor. It is weaponized sound in its purest form.

3. "Dark Was the Night, Cold Was the Ground" – Blind Willie Johnson

Context: Chapter 11. There are no lyrics, only a moan and a slide guitar. It is the definitive example of metabolizing misery. It was chosen for the Voyager Golden Record to represent "human loneliness" to aliens.

4. "Keening Song (Caoineadh na Marbh)" – Traditional Irish

Context: Chapter 11. Listen to the melodic wail. It is not pretty; it is functional grief processing.

5. "Hurt" – Johnny Cash

Context: Chapter 15. The sound of a dying voice. The anti-Auto-Tune.

6. "Gymnopédie No. 1" – Erik Satie (Humming Experiment)

Context: Chapter 9. Use this slow, simple melody to practice the "Vagal Hum." Close your mouth, drop your jaw, and hum the melody in your chest.

CITATIONS NOTE

Every study referenced regarding heart rate variability, cortisol reduction, oxytocin levels, and pain tolerance is drawn from peer-reviewed journals in the fields of music psychology, neurobiology, and sports science. While this book is written as a manifesto, the mechanics are peer-reviewed.

www.ingramcontent.com/pod-product-compliance
Lightning Source LLC
LaVergne TN
LVHW022222200926
848481LV00006B/2076